The Expression of Time

The Expression of Cognitive Categories

ECC 3

Editors
Wolfgang Klein
Stephen Levinson

Mouton de Gruyter
Berlin · New York

The Expression of Time

edited by
Wolfgang Klein
Ping Li

Mouton de Gruyter
Berlin · New York

Mouton de Gruyter (formerly Mouton, The Hague)
is a Division of Walter de Gruyter GmbH & Co. KG, Berlin.

♾ Printed on acid-free paper which falls within the guidelines
of the ANSI to ensure permanence and durability.

Library of Congress Cataloging-in-Publication Data

> The expression of time / edited by Wolfgang Klein, Ping Li.
>   p. cm. − (The expression of cognitive categories ; 3)
>   Includes bibliographical references and index.
>   ISBN 978-3-11-019581-1 (hardcover : alk. paper)
>   ISBN 978-3-11-019582-8 (pbk. : alk. paper)
>   1. Grammar, Comparative and general − Temporal constructions.
>   2. Grammar, Comparative and general − Tense.   3. Tense (Logic)
>   4. Time.   I. Klein, Wolfgang, 1946 Feb. 3−   II. Li, Ping, 1962−
>   P294.5.E97   2009
>   415−dc22
>   2009004485

Bibliographic information published by the Deutsche Nationalbibliothek

The Deutsche Nationalbibliothek lists this publication in the Deutsche Nationalbibliografie;
detailed bibliographic data are available in the Internet at http://dnb.d-nb.de.

ISBN 978-3-11-019581-1 hb

ISBN 978-3-11-019582-8 pb

© Copyright 2009 by Walter de Gruyter GmbH & Co. KG, D-10785 Berlin.
All rights reserved, including those of translation into foreign languages. No part of this
book may be reproduced in any form or by any means, electronic or mechanical, including
photocopy, recording, or any information storage and retrieval system, without permission
in writing from the publisher.
Cover design: Frank Benno Junghanns, Berlin.
Printed in Germany.

Contents

Introduction 1
Wolfgang Klein and Ping Li

Concepts of time.. 5
Wolfgang Klein

How time is encoded.. 39
Wolfgang Klein

Temporal anaphora in a tenseless language: the case of Yucatec........ 83
Jürgen Bohnemeyer

Tenses in compositional semantics................................ 129
Arnim von Stechow

Temporality in first and second language acquisition.................. 167
Yasuhiro Shirai

New perspectives in analyzing aspectual distinctions across
languages 195
Christiane von Stutterheim, Mary Carroll and Wolfgang Klein

Verb aspect and the mental representation of situations 217
Carol J. Madden and Todd R. Ferretti

Computational modeling of the expression of time 241
Ping Li and Xiaowei Zhao

Contributors... 273
Index.. 275

Introduction

Wolfgang Klein and Ping Li

> There is no present or future, only the past, happening over and over again, now.
>
> Eugene O'Neill

The ability to express time belongs to the most fundamental traits of human communication. All human languages that we know of provide their speakers with a range of lexical and grammatical devices to say when something happened and how long it lasted, to say whether it happened, or will happen, for the first time, regularly or very often, and to say whether some event or state precedes, overlaps with or follows another event or state. These devices include grammatical categories such as tense and aspect, certain features in the lexical meaning of verbs, various types of temporal adverbials and particles, but also discourse principles such as the maxim to tell events in the order in which they happened.

In many languages, one of these devices, tense, is so deeply rooted in the grammatical system that it is hardly possible to utter a sentence without referring to time. This may be the reason why the study of how time is expressed in human languages has a strong bias towards tense and, to a somewhat lesser extent, aspect – two categories which are often combined to what traditional grammars usually call a "tense system". This bias is unfortunate for two reasons. First, many languages have no inflectional morphology at all, hence, no categories as tense and aspect in their grammatical system (e.g., Chinese has no grammatical tense). This does not mean, of course, that the speakers of those languages cannot indicate that something is in the past, the present, or the future, or that something is on-going or completed; they just use other means, for example particles or adverbials. Second, in languages like Greek, English or French, in which tense and aspect are parts of the grammatical system, the expression of time is not confined to these two devices. Typically, the temporal information which the speaker wants to convey is encoded by a combination of various means, including adverbials, inherent temporal features of the verb and discourse principles. Hence, any real understanding of how the expression of time works requires a somewhat broader perspective. This book tries to provide such a perspective.

Our book is not meant to be a comprehensive survey of what linguistics and its neighbouring disciplines have found out about the various facets of how time is reflected in human languages. Given the wealth of literature on this issue, this would be a hopeless task. The book should rather be considered as an invitation to study how time is expressed in human language, and to provide a good starting point for such an enterprise. Its chapters take the reader through a number of foundational issues, such as the various notions of time and the various devices found in different languages; other chapters are devoted to more specific questions, such as the acquisition of time, its modelling in formal semantics and in computational linguistics, or how its expression can be empirically investigated. In this way, it reflects the state of the art, on the one hand, and it also aims to pave the way for future research, on the other.

Time is not a uniform notion. There is the time of physics, the time of biology, the time of psychology, the time of economy; there is the time of philosophers, of anthropologists, of linguists, of orchestra conductors. All of these notions share certain characteristics, and they differ in others. In the first chapter "Concepts of time", *Wolfgang Klein* briefly characterizes a selection of these concepts; he then discusses three perennial issues in the study of time since the Antiquity; these are the relation between "time and change", the "units of time", and the role which the "now" of the observer plays in concepts of time. In the last section, he sketches the core components of a "basic time structure" which underlies the expression of time in human languages.

The next chapter "How time is encoded" by *Wolfgang Klein* examines six main devices which human languages use to express time; tense, (grammatical) aspect, "Aktionsarten" (lexical aspect), temporal adverbials, particles, and discourse principles. By far most research on temporality is devoted to the first three of these devices. It is shown that the established definitions (for example "tense is a grammatical category which serves to localise the time of event in relation to the moment of speech" or "aspect is a grammatical category which serves to present the event as on-going or completed") raise a number of non-trivial problems. The chapter concludes with a brief discussion of how these problems might be overcome.

The third chapter, *Jürgen Bohnemeyer's* "Temporal anaphora in a tenseless language: the case of Yucatec", is an in-depth investigation of a Mayan language which lacks deictic tense (in the usual understanding) as well as words which correspond to English *before, after, while, until*. Still, its speakers are no less able to make appointments and to tell stories with a complex temporal structure than speakers of English or French. Bohnemeyer demonstrates that their system is based on a very systematic and clever use of various types of temporal anaphora. His analysis not only shows that temporality can function very differently from what we are used to think; it also challenges us to have a fresh look at those languages which are supposed to be relatively well-described.

Whereas all natural languages have temporal expressions, formal languages – for example languages of logic or programming languages – typically lack such devices. But about forty years ago, philosophers and linguists began to develop more complex formal systems which permit a precise analysis of how, in natural languages, the meaning of compound expressions results from the meaning of its components. *Arnim von Stechow's* chapter "Tenses in compositional semantics" illustrates this for English; beyond tenses proper, he also examines lexical and grammatical aspect.

We do not know in which way "human time" – that is, the way in which we experience time and think about time – is shaped by our genetic endowment, on the one hand, and by social and cultural experience, on the other. But clearly, the way in which temporality is expressed must be learned, since languages differ considerably in this regard, no matter how similar the underlying temporal notions may be. In his chapter "Temporal expressions in first and second language acquisition", *Yasuhiro Shirai* gives a survey of what we know about these developmental processes, in particular how the acquisition of the mother tongue differs from the acquisition of a second or third language.

One reason why the study of temporality has made much less progress than it could have done is a certain methodological narrowness, coupled with an unbalanced diet of examples: by far most claims in the literature are based on speaker's intuitions of what certain forms – say *he was sleeping* vs. *he slept* – mean, or on the study of occurrences of such forms in a text corpus; in either case, the data have a strong bias towards utterances which describe singular events in (real or fictituous) past. Both procedures, even if extended to other text types and usages, have serious shortcomings. The chapter "New perspectives in analysing aspectual distinctions across languages" by *Christiane von Stutterheim, Mary Carroll* and *Wolfgang Klein* shows that other – and actually very simple – experimental methods can give a much more reliable and differentiated picture of how speakers of different languages use certain temporal forms. This is illustrated here for aspect; but the methods can easily be extended to other temporal devices.

There is a clear difference between the temporal properties of a situation (state, process, event) itself and the mental representation which the speaker has of this situation, when he or she sets out to speak about it. It is the speaker's mental representation, rather than the situation itself, which is crucial for the linguistic expression of time. The more or less subjective "view" on the situation is closely connected to grammatical and lexical aspect. In the chapter "Verb aspect and the mental respresentation of situations", *Carol Madden* and *Todd Ferretti* first examine the traditional classification of these two notions, and then, they present evidence from recent psycholinguistic experiments on how speakers construct the discourse model that underlies their language production.

The concluding chapter "Computational modeling of the expression of time" by *Ping Li and Xiaowei Zhao* is devoted to a very different way to investigate temporality – computational approaches. The strong association and interaction between lexical aspect and grammatical aspect, particularly in the domain of first language acquisition, has previously led some researchers to argue for innate semantic categories or prelinguistic predispositions. Ping Li and Xiaowei Zhao provide counter-evidence to this argument with simulations of aspect acquisition in a connectionist network, DevLex-II. The simulation results indicate that the strong association between lexical aspect and grammatical aspect can emerge from dynamic self-organization and Hebbian learning in dynamic computational systems, therefore invalidating a priori assumptions about specific structures of innate linguistic or conceptual knowledge. Thus, computational modeling is not an aim in itself – it is a tool to verify or falsify existing claims on how time is expressed, and to provide novel simulation data that can further inspire new empirical studies.

The expression of time marks an important human linguistic capacity, and the study of it requires the joint efforts of linguists, psychologists, and other cognitive scientists. It is our hope that this book serves as a catalyst for future research that will elucidate the cognitive and linguistic processes underlying the expression of time.

Concepts of time

Wolfgang Klein

> Time has always been difficult to understand, but in the twentieth century, our ununderstanding has become clearer.
>
> J. R. Lucas (1999: 1)

1. Introduction

The experience of time and the need to adapt our life to it are as old as mankind. The sun rises in the morning and sets in the evening, the moon changes its position at regular intervals, plants and animals and humans come into existence, grow, fade and pass away. We act here and now, but we also remember having acted, and we plan and hope to act in the days ahead of us. Some of these events, such as the coming and going of the seasons, are cyclic, that is, they are repeated at intervals which we consider to be equal. Other events are not assumed to be cyclic, such as our first love, the birth of Jesus, or Grandmother's death. All human cultures and societies of whom we know have reacted in three ways to this temporal nature of experience:

- First, *actions are planned and done* accordingly – there is a time to plant and a time to reap; a time to tear down, and a time to build; a time to mourn, and a time to dance, as the Preacher has it in the Bible.
- Second, methods to *measure time* were invented. This is always done by linking some event – the event whose duration we want to measure – to some other type of events which are supposed to occur at regular intervals, such the sequence of the seasons, the fall and rise of the sun, the swing of a pendulum, the oscillation of a quartz crystal; the result are calendars and clocks (Bruton 1993; Landes 1983; Richards 1998).
- Third, we *speak* about time. All human languages have developed numerous devices to this end, and in some languages, the marking of time is even close to mandatory. In English, as in all Indoeuropean languages, the finite verb regularly expresses "tense" – that is, the sentence not only describes some event, process, or state. It also places this situation into the past, present, or future: we cannot say *John be ill*, thus leaving neutral the time of the state thus described. We must say *John was ill, John*

is ill, John will be ill. Other languages, such as Chinese, have no mandatory marking of time. But this, of course, does not mean that they cannot express time; they just use other means, such as adverbials like *yesterday, right now,* or *very soon,* and they give their speakers full freedom to indicate what happened when.

So, we all adapt our life to time; we use devices by which time is counted and measured; and, above all, we speak about time. We know what it means when someone says *He will arrive tomorrow at five., The meeting has now lasted for almost eleven hours.,* and *Last february, I intended for the first time to spend more than three hours per week in Pontefract.* So, we do understand what time is. But what is it, then?

At this point it is common to quote St. Augustine, who, in the 11th book of his *Confessions*, says:

Quid ergo est tempus? Si nemo ex me quaeret, scio. Si quaerenti explicare velim, nescio.
[What, then, is time? If nobody asks me, I know. If I should explain it to someone who asks, I don't know].

His own way to overcome this clash between practical and theoretical understanding of time is that time is not in the things themselves but in our soul. God, he says, is beyond time, and we get to know all things created by him because he has endowed our soul with memory, experience, and expectation (see Flasch 2004). In other words, our soul – or our mind, as we would probably say now –, is such that we experience the world as *past, present,* and *future.*

St. Augustine's theory of time is one of many within a rich stream of thought that began with the first Greek philosophers and has steadily unfolded over the millenia and over many disciplines – philosopy, physics, biology, psychology, anthropology, linguistics, to mention but these. They all deal partly with the same and partly with different aspects of time, the result being a hardly permeable jungle of views, opinions, and theories. In fact, the Augustinean question "Quid ergo est tempus?" has found so many answers that one might as well say that there is no answer at all. Thus, the idea that we could ever grasp "the essence of time" is perhaps futile; is is doubtful that there is much more than a kind of family resemblance between a biologist's, a phycisist's, and a psychologist's concept of time.

The aim of this chapter is not to unveil the "very nature" of time; it is rather to prepare the ground for a basic understanding of how temporality is expressed in natural languages.[1] To this end, it is necessary to gain some idea (a) of the underlying temporal notions, and thus of what people understand by "time", and (b) of the means by which these notions are encoded in the different languages of the world. The second issue is addressed in the following chapter of this book. The present chapter is devoted to the notional category of time. Section 2 reviews the diversity of meanings with which this category is associated; we will glance at some of the key questions which are dealt with in different fields. Section 3 discusses three perennial issues which come up time and again when people reason about time. In section 4, I will sketch a "basic time structure", which, I believe, can serve as a useful starting point for the study of how time is expressed in language.

The following exposition is strongly biased towards "the Western tradition" of reflection on time. Apart from shere lack of knowlege on my part, this bias has three reasons. First, it is by far the best studied tradition; there is, of course, research on non-European notions of time; but it is comparatively sparse (see, e.g., Needham 1968; Fraser, Haber and Lawrence 1986). Second, only in that tradition do we find this enormous spread of temporal notions across various disciplines, such as physics, biology, or psychology. Third, different as human languages are – our entire way of thinking about their lexical and structural properties is deeply shaped by the Western tradition of linguistics. In Latin, the word *tempus* means both "time" and "tense", and thus, one is easily led to believe that tense is the most immediate reflection of time in language, in fact, that tense is time. This close connection has misled not only linguists but also philosophers who think about time, and so, it is important to understand its roots.

2. The variety of time

This section is a gaze into a jungle – into the rank growth of notions, ideas, problems which have grown from a few germs laid in the Antiquity. At first glance, it would appear to be hopeless to detect any structure in this jungle; but in fact, there are a few recurrent themes which we will address in the following section. It should be clear that this panorama is anything but ex-

---

[1] The number of books on time is legion. The best general survey is to my mind Whitrow 1980; it is, however, confined to time in philosophy, physics and biology. Fraser 1987 is an easy and broad introduction by one of the best experts on the study of time.

haustive: it is simply meant to give an impression of the abundance of temporal phenomena.

We will begin with philosophy – the mother of any science and the origin of human thought on time. In fact, any such reasoning reflects a particular perspective on reality and the way in which we are able to recognize it – a perspective on us and on the world around us. In this sense, any reflection on time is inevitably "philosphical". But if we speak of "philosophical" theories of time, in contrast to, for example, physical theories, we normally mean the more or less elaborate views of particular philosophers, from Anaximander[2] to Heidegger and Wittgenstein. There are many such theories; here are three characteristic examples from modern times; they stand for very different perspectives on time (Turetzy 1998 is an excellent survey; Le Poidevin and McBeath 1993 is a characteristic collection of articles on 20th century philosophy of time).

A. In Immanuel Kant's *Critique of pure reason* (1781), time and space are properties of human cognition – in fact, the two most fundamental categories of human cognition. They define the way in which our mind experiences, and thinks about, the world. Time, in particular, is "die innere Form der Anschauung", (the inner form of intuition). It defines the way in which we "intuit" external events and facts, such as the running of a horse or the rotation of the earth, but also internal events, such the feeling of hunger or grief. We cannot know whether time is "real", that is, a property of the world itself; our cognitive apparatus is such that the outer as well as the inner world inevitably *appear* to us as structured by time.

B. In his influential article *The unreality of time* (1908), the British philosopher John McTaggart argued that there are two types of event series, each of which represents time: The "A-series" relates to the "earlier-later"-order, to the mere *succession* of events, states, processes. In this sense, Aristotle lived before St. Augustine, and Kant lived after St. Augustine. The "B-series" relates to the difference between "past – present – future". In contrast to the A-series, it requires a particular vantage point, from which the events are seen; this is the present moment – which, in turn, permanently shifts. Under neither understanding is time "real", argues McTaggart (see, e.g., Turetzky 1998).

---

[2] In what is probably the oldest fragment of Greek philosophy we have, Anaximander says that the things, as they come into existence and perish, "pay their debts to each other according to the order of time" ("kata tou chronou taxin") – a sentence of which no element is easily understandable (see Turetzky 1998: 6–8).

Kant's and McTaggart's views on time are among the most-discussed in modern philosophical literature; but they are not really new – they elaborate and extend themes that are already found in the antiquity. As we have seen above, St. Augustine also thought that time is a property of our "soul", and that it divides the world, as we can recognise it, into past, present, and future; he also had a clear notion of succession being a crucial feature of time. This is quite different for the third philosophical theory of time which I will mention here.

C. In Martin Heidegger's book *Sein und Zeit* ("Being and Time"), published in 1927, time is not so very much seen as an objective property of the world around us or a subjective property of our way to know the world in or around us. Rather, it is something that shapes human existence. Human time is not an abstract order, real or imaginary, defined by relations such as "earlier" or "simultaneous". It is the slope which separates us from death, a short stretch filled with our sorrows and efforts and griefs. It is the notion of time which surfaces in expression such as "little time is left", "these were hard times", or, in the words of the Preacher quoted above, "there is a time to plant and a time to reap; a time to mourn, and a time to dance". Such a notion of time is not incompatible with the idea of succession and the division into past, present and future; but properties like these are somehow marginal to what time means for humans.

These are three of the very many ways in which philosophers have looked at time. They may not be mutually exclusive; it is not even clear whether they target the same entity or not. And in a way, we do not expect the opinions of philosophers to converge on some phenomenon. But we do expect this in hard science. So, there should be one notion of time in physics. This is not the case. There are at least three approaches towards this chimera.

D. The first of these is the view which underlies the laws of classical mechanics, as first stated by Isaac Newton. In the introductory "definitions" to the *Principia*, Newton distinguishes two notions of time:

> Tempus Absolutum, verum & mathematicum, in se & natura sua absque relatione ad externum quodvis, æquabiliter fluit, alioque nomine dicitur Duratio: Relativum, apparens, & vulgare est sensibilis & externa quævis Durationis per motum mensura (seu accurata seu inæquabilis) qua vulgus vice veri temporis utitur; ut Hora, Dies, Mensis, Annus.
> (Newton, Principia, Book I, Scholium to the Definitions)
> ['Absolute, true, and mathematical time, in itself, by its very nature and unrelated to anything external, flows equably, and is also named Duration: relative, apparent, and everyday time is some sensory and external (accurate

or unequable) measure of duration by motion, and it is is commonly used instead of true time; such as hour, day, month, year.']

A number of points are remarkable in this short paragraph:

(a) Time is the same as Duration. It is neither an order, defined by "earlier" or similar notions; nor is it in any way related to past, present, future. This does not mean that Newton had no idea of *succession*; in a somewhat mysterious way, it comes in in the term "aequabiliter fluit". But in its absolute as well as in its relative understanding, Newton equates time with *duration*.

(b) We cannot measure "real" time – whatever it is. Instead, we measure the duration of things to which our senses have access. This duration is "relative time"; it is measured by motion, and the result are units such as hour, day, year, etc.

(c) Real time is always the same; still, it "flows", and it flows equably – whatever that means. Newton does not say whether it flows in one direction, although this would seem the most natural assumption. Real time is, so to speak, unaffected and unaffectable by anything. In fact, it is not even related to anything "external"; in particular, it is not related to any *observer*.

Newton's notion of real time is cryptic, perhaps because it has a strong religious background. As he states in the Scholium Generale of the second edition of the Principia (1713) – an addition which particular famous for Newton' statement "Hypotheses non fingo (I don't make up hypotheses)" – he argues that time is an emanation of God, and God is in time (a position which is in sharp contrast to St. Augustine's, according to whom God is out of time). It may well be that the tremendous success of Newtonian mechanics is completely independent of his conception of "real" time. What is crucial for the laws of motion is the possibility of measuring the time of observable events by motion. This is not possible for real time. What really matters in Newtonian physics is thus relative time. Absolute time, dear as it may have seemed to Newton, is something that lurks in the background, and is perhaps completely superfluous to the physicist.[3]

---

[3] His great opponent Gottfried Wilhelm Leibniz argued that time and space are purely relational – there is no absolute time and no absolute space. In response to this, Newton's spokesman Samuel Clarke gave an argument as to why we need something like "empty space", independent of the properties of objects that are "in space". But no corresponding argument was ever given for "empty time" (see Westfall 1983).

E. Classical physics, including its notion of time as duration, sometimes leads to undesirable asymmetries. If, for example, a conductor and a magnet move in relation to each other, then there is a electromagnetic effect. This effect should be the same no matter whether the conductor moves, or the magnet moves. But classical physics gives two completely different accounts of the effect for both cases. The problem disappears if we assume that there is no "distinguished frame of reference", in particular no absolute frame of reference, as seems to be implied by absolute time and space. We can choose the position of the conductor as well as the position of the magnet as frame of reference; the laws of physics operate in the same way, no matter what the frame of reference is; the only factor that remains constant is the velocity of the light. This is the basic idea which Albert Einstein worked out in 1905 in what was later called "special theory of relativity" and which, among others, led to the notion of "relativistic time". This time has peculiar properties, which are often felt to be paradoxical; thus, it may shrink or extend – an idea which seems very different from a notion of time which flows equally and is unaffected by anything external, and no less different from our everyday notion of time.

Usually, a sharp constrast is made between Newton's absolute time and Einstein's relativistic time. This is misleading, because in actual fact, Newtonian physics does not operate with absolute time, either. Absolute duration (scil. absolute time) exists, but it is not accessible to us; all we can measure is relative time. What Newton did not consider was the possibility that the measurable duration of some event could vary with a frame of reference ("Koordinatensystem", as Einstein says in German); exactly this assumption is made in Einstein "relativistic time". But Newton never spelled out how relative duration differs from absolute duration, except that the former is a familiar phenomenon and can be measured by motion, whereas the latter is the "true" duration.

Relativistic time and Newtonian time (in both variants) have three properties in common:

(a) What is crucial, is not so much the "earlier – later" order of observable phenomena – their *succession*; it is their *duration*. The famous and perplexing "time dilation" and "time contraction" effects of the special theory of relativity refer in the first place to changes in the duration of some observable phenomena, when measured from different frames of reference. But indirectly, varying duration also affects observed simultaneity and succession between two events. The reason is that *information* about these events needs some time to reach the observer, and this

time takes longer or shorter, depending on the relative distance between the place where the events occur, and the frame of reference.
(b) The laws of physics operate equally "in both directions". They do not go "from earlier to later" or "from later to earlier". This asymmetry, so fundamental to the daily experience of time, plays no role under these two conceptions of time.
(c) Similarly, the observer – the person who experiences time – plays no role. There is no past, present or future, no shifting Ego, in relation to which these notions are defined. In relativistic time, there is always a "frame of reference"; but all that is relativised is duration. Einstein never denied that past, present and future are important ingredients of everyday notions of time; but not so in the world of physics (cf. section 3.3 below).

Since the laws of physics do not conform to an "arrow of time", which invariably flies from earlier to later, they reveal the kind of symmetry which physicists like; the theory of special relativity started as an attempt to overcome an undesirable asymmetry. But has nature really no earlier-later orientation? Is the real world, whose laws the physicists try to find, like that? Questions of this sort have given rise do a different notion of physical time.[4]

E. We can imagine that an egg, once fried, returns to its initial state; we can even have a film run backwards, thus apparently reversing the order of time. But we never observe such a return in reality. There are many physical processes which, it seems, obey the "arrow of time". A well-known type of such unidirectional processes are the changes of entropy (roughly: the amount of disorder) in a closed system, as studied in thermodynamics. In Clausius' formulation from 1851, the second Law of Thermodynamics states that the overall entropy (roughly: the amount of disorder) of a closed physical system can remain constant or it can increase; but it cannot decrease, unless such a change is caused by influences from outside the system: inherent state changes of the entire system are unidirectional. This has given rise to a physical concept of irreversible time, a concept which is neither Newtonian nor Einsteinian (see, for example, Prigogine and Stengers 1993). It should be noted, though, that irrreversibility is not to be equated with the earlier-later asymmetry, as is often done. Even if the fried egg could be re-

---

[4] Reichenbach (1958) is still a very clear treatment of this problem; see also Horwich (1987) and the contributions in Savitt (1993) for a more recent discussion.

stored, the time at which at which it has its original shape again is *later* than the time at which it was not yet fried: the egg is as it was *before*. We will come back to this problem in section 3.1.

The time of physics, in whichever of the three variants mentioned here, does not integrate some of the features which we normally associate with time. It deals with the temporal structure of dead matter, not of living organisms. There are at least three notions of *biological time* – the life span of the individual, biological evolution, and biological rhythms in the organism.

F. The life of an indivdual has a beginning: birth (or perhaps conception). It has an end: death. And the processes between birth and death are, as a rule, not reversible: they have a certain duration, and they are fundamentally characterised by the earlier and later of growth and decay. This second fact makes biological time crucially different from physical time in the Newtonian or Einsteinian sense. It makes it also different the time notion of thermodynamics; there is no organic "growth and decay" in the changes of closed systems, except in a very metaphorical sense.

G. Antique and mediaeval thought did not consider the world as entirely static. There are changes, such as the motion of bodies, the changing seasons, or even the notion of subsequent ages – for example, a "Golden Age" followed by a "Silver Age". But it was not until the late 18th century that the idea of *evolution* gained ground – that is, of a temporally directed and rule-governed process which determines directed changes of whole systems, usually towards an increasing complexification. The earliest detailed treatment I am aware of is by Johann Gottfried Herder (1784, vol. I). Such systems might be, for example, *languages*; Hermann Paul, one of the leading linguists of the 19th century, even argued that only historical linguistics deserves the name of a science, because only this way of looking at language reveals the principles that underly it, rather than merely stating facts Paul 1882: 20). They might be *physical systems*, such as the earth, the solar system, or even the entire universe. But by far the most discussed example is the origin and evolution of *life*, which, as is now generally believed in the educated world, is determined by a few principles such as genetic variation, extinction according to fitness, or drift.

H. There is a third way in which we can speak of biological time. Many processes within a living organism follow a "biological rhythm", for example the circadian rhythm which, as a rule, lasts 24 hours in human beings, though with considerable variation. These rhythms are essentially determined by a "timer" – maybe several timers – inherent to the organism, but this internal timing interacts with influences from outside the organisms, for example the amount of light or heat. This highly complex and only partly understood interaction regulates order, duration and intensity of physiological processes in the organism (a good survey of the present state of research is given in Foster and Kreiman 2004).

In biological research, these rhythms are usually characterised in terms of chemical processes in various types of organisms, flowers, animals, human beings. But they bring us already somewhat closer to the properties of a person who actually experiences time – a notion completely absent in physical time. We find it, for example, in St. Augustine's notion of time as a property of our soul. His argument is entirely based on a very subtle but completely intuitive self-observation. Modern psychology has led to many insights into how time is perceived, remembered and transformed into human actions.

I. What is the *Now* that separates the past from the future and allows us to define what is present? This notion has vexed philosophers from Aristotle to McTaggart, for at least two reasons. First, it "shifts" permanently: there is not a single now, there are infinitely many nows. But there is always a special now – the now right now, so to speak. So, how is this now defined in contrast to all the other nows? Second, the "now" is supposed to have no extension, hence no duration (and in the sense of physical time, it is not in time at all: no duration, no time). If this is true, then there can be no present. But if there is no present, it seems to make little sense to speak of past and future. Arguments of this sort have lead to the idea that time is "not real", a position indeed taken by philosophers from the antiquity until to McTaggart. Now, rather than worrying about these puzzles, psychologists have tried to determine what the minimal unit of perception is, that is, the shortest time at which our sensory organs can, for example, distinguish a change in vision or audition. For human beings,[5] this shortest moment is assumed to be somewhere between 30 and 40 milliseconds (Poeppel 1988). But already William Stern noted in 1897 that this shortest moment may not coincide with what we consider to be "present" (whose duration he calcu-

---

[5] The idea of such a shortest time span of perception and the possibility that it might vary across species was first introduced the founding father of embryology, Ernst Baer, in 1864. He also beautifully illustrated the dramatic consequences of this variation for the way in which the world is experienced.

lated as 6–7 seconds). This may not solve the philosophical problems connected to the now; but if the present is defined by what is perceived right now, then we know at least how long the now is.

J. How do we experience duration? The duration of some event or state is "objectively" measured by relating it to repeated occurrences of some other event (for example the heart beat or the rotation of the earth around its axis). As everybody knows, this measured duration of an event often sharply contrasts with the subjective duration someone attributes to it. This variation depends on many factors, for example

- the number of subevents – that is, changes we note within the entire period: if "nothing happens" within one hour, then this hour is subjectively much longer than when it is filled by many subevents;
- the degree to which we like the event: sadly, unwanted events seem to last much longer than events which we enjoy;
- the influence of drugs; some drugs "stretch" the subjective duration of an event.

We do not immediately perceive the relative order of events – succession always involves memory or expectation. This brings us to the second factor, the memory of time.

K. Time is closely connected to remembrance. But how do we remember time? This concerns duration as well as succession. In our recollection, the perceived duration of an event is sometimes reverted: idle hours, which did not seem to end, shrink in memory, events which excited our attention and seemed to pass rapidly, as they happened, tend to be very long in memory. If we try to recollect a complex event that we have experienced in the past – say a car accident –, how do we know that subevent A came before subevent B? And how do we record partial overlaps of subevents? In other words, how do we store the order of events, as we perceived them? In some cases, we might have looked at a watch and thus remember "A was at 10:15, B was at 10:16"; but this is surely the exception. Do we use an inner watch which allows us to stick a sort of "time stamp" on all subevents? Or do we just associate pairs of events by a relation "A before B" or "A simultaneous with B", thus eventually forming a complex temporal web of subevents? (see Kelly 2005 for a discussion and how it relates to various other time puzzles).

L. We not only perceive and remember temporal features of what happens in our environment, we also plan and perform actions. These actions often consist of a complex structure of simultaneos and sequential subactions. Thus, they exhibit a complex temporal structure. In some cases, the temporal order in which subactions are to be performed are more or less dictated by the intended result ("first the socks, then the shoes!"); in other cases, this must be stored as an independent part of the planning ("first push the red button, then the black button"!). Jean Piaget, in his famous theory of child development (1927), argued that a great deal of this development is characterised by increasing abilities to decompose complex actions in subparts and to process and execute them independently; young children treat complex events holistically, as a unit, older children learn to separate and possibly revert its parts. A particularly interesting aspect of the temporal composition of actions concerns the question of whether our "decision" to do something always precedes the action itself. One such subphase of a action concerns the decision to perform it: does this decision always precede the action itself? Benjamin Libet and others have shown that this may not always be the case – a finding which has led to considerable discussion on the notion of a free will (Libet 2004).

Humans are similar in some respects, and they are different in others. To what extent does this influence their notions of time? No one assumes that biological differences between individuals bear on the relative order of events, or the division of time into past, present and future. This is perhaps less true for duration: some people seem to be slow, others are fast, and this could be due to the fact that their inner clock runs at different speed. A good example is language processing; there is a number of verbal tasks in which women are on the average much faster. But this biological variation is minor, when compared to the variation in human cultures. In anthropological research, it is often assumed that different societies have developed quite different concepts of time. In what follows, we briefly discuss four examples which illustrate this variation (a very detailed discussion is found in Wendorff 1980).

M. Life in different cultures always follows certain "natural rhythms", such as the sequence of the seasons or the various ages of a person from birth to death. But the degree to which these rhythms dictate human life and thought varies considerably. Societies in which these rhythms prevail are often said to prefer a "cyclic concept" of time, in contrast to the "linear time", so familiar to us in modern Western societies.

N. A second, related aspect is the degree to which daily life and work are dictated by the mechanical measurement of time. Until a few centuries ago, precise clocks and calendars were exceptional in any society; nowadays, they characterise the entire daily life of more and more societies (see, for example, Dohrn-van Rossum 1996). Note, however, that "mechanical clock time" is not to be equated with the notion of linear time; after all, clocks are based on cyclic events.

O. What role do history and chronology play in a society? All human cultures we know of have some notion of the forebears which may still be "present" in some sense – dead, but still an active force in daily life. This connection to the past may be structured in different ways. Some old cultures, such as the Chinese, the Japanese or the Egyptian culture, are bound to the remote past by an uninterrupted "chain of generations", for example dynasties or families. Others, such as the Greek or the Indian culture, also have strong ties to a very distant past, but they never had the notion of such a linear chain which connects the present to the origin (see Nakayama 1968).

P. Cultural variation in timing also surfaces in a number of phenomena which we find in all human societies. The most obvious example is *music*: in all its manifold forms, it is always a way to organise sounds in time – to organise their succession or simultaneity, as well as their duration. Music has its roots in our biological clocks; but the way in which it evolves varies massively across cultures (Jourdain 1999). Many other time-related human activities show the same pattern: there is an essentially universal biological root, and there is massive cultural variation – for example dance, poetry, and, of course, language.[6]

This brings us to our final point. There are at least four ways in which language is crucially connected to time. Languages *change in time*, they are *processed in time*, they exhibit a *linear order*, and they *express time*.

Q. In the antiquity and in the middle ages, the idea that languages change was not unknown; but this fact, obvious as it is, did not play a substantial role in the way in which philosophers and linguists thought about language.

---

[6] Another interesting case are movies which present a complex event within a certain time frame, say 90 minutes; but the "real" time of the event thus represented is, of course, normally much longer. This can be used for special effects, such as in Buster Keaton's silent movie "Seven chances" in which movie time and depicted time converge, as the movie goes on.

This changed with the advent of historical comparative linguistics around 1800, and for at least a century, diachronic research reigned in the study of human languages. This research has bestowed a tremendous amount of empirical facts upon us, albeit only for a small number of languages. But in contrast to biological evolution, we are still very far from an idea of the principles that determine how human languages change over time (a good survey of the state of the art is Janda and Joseph 2004).

R. One of the miracles of human language is the speed with which it is processed in everyday communication. This becomes immediately clear as we look at a simple question-answer sequence such as: "Where were you born? – In Heidelberg". The person who answers has to identify the sounds, words and rules of the question in about one second, and it takes about another second to produce the answer. This includes the repeated inspection of something like 50,000 lexical items somewhere stored in the brain, but also the storage of the syntactic pattern of the question, the decision to use this pattern in the answer and to omit those parts which would be identical (the answer means "I was born in Heidelberg", and not just "in Heidelberg"), the innervation of a complex articulation pattern, etc etc. (Dietrich 2007 gives an excellent survey of this research)

S. There are three major modalities by which human languages are encoded – speaking, writing, gesturing. Each utterance, each text follows a *linear order*, which is fundamentally temporal in nature. Linguists often say that a constituent "is moved to the left" or "to the right"; but in fact, this is only a spatial metaphor for the fact that this constituent is somehow processed at an earlier or later time when pronounced or written, heard or read.

T. Independent of whether a culture has a more or less elaborate theory of time – its members are always able to speak about time. They relate personal experiences, they talk about their future plans, they arrange dates, they describe how to bake a cake – all of this requires temporal notions of duration, succession and simultaneity. For a long time, the study of how temporality is encoded was completely dominated by two grammatical categories, tense and aspect, and a lexical category, called Aktionsart, situation type, or sometimes lexical aspect (Binnick 1990 gives an excellent survey of this research tradition). But this is only a selection of the means which natural languages use to express time; temporal adverbials are by far the most elaborate means. These devices will be discussed in the following chapter of this book.

This concludes our panopticum of time; it is easy to see that it is anything but exhaustive; but it surely suffices to give a picture of the diversity of time. Let us return to the initial question: What, actually, is time? If anything is clear by now, then it should be the fact there that is not a single notion of time. But it is also clear that the many facets of time are not an arbitrary collection of phenomena. There are a number of discernable threads in this clew, three of which will be addressed in the following section.

3. Three recurrent themes

3.1. Time and change

There is no immediate experience of time itself. What we experience are changes around us and in us. We see that it is getting dark and that it is getting light, we feel cheerful, and we feel sad. This experience is ubiquitous, and people have to adopt their life to it. But it was only the early Greek philosophers that began to wonder about two things. The first of these is a fundamental ontological problem: what does the pervasive experience of change tell us about the nature of reality:

- Is it steadily changing, as Herakleites is supposed to have thought? Among his cryptic sayings, *panta rhei* "everything is in flow" is probably the most famous.
- Is this impression of steady change fallacious, and reality is eternal and immutable, as Parmenides is reported to have thought?
- Do we have different types, or perhaps degrees, of reality – one of them characterised by change, the other one by non-change, as Plato and many others, notably the Neo-platonists, argued?

In this discussion of reality, two issues must be clearly kept apart: The first issue concerns the "reality of time": is time "real", or is it just a fiction of our mind. This question has led to vivid discussions, but mainly among philosopers; it is an interesting but somewhat academic problem. The second issue is the nature of reality itself: is there a reality – maybe the only "real reality" – behind the apparent changes, which our senses tell us? Different views which people have taken on this question had dramatic consequences in the history of mankind; the entire dogma of transsubstantiation, so fundamental to Christian faith, depends on the possibility that "real reality" is independent of apparent change or non-change, and many people have died for the one position or the other; so, it is probably an important issue.

The second problem, about which the Antique philosophers stumbled, is not ontological but epistemological. How is it possible that one and the same entity can have two mutually incompatible properties? How can someone be alive and dead, how can someone be in Athens and in Crete? The answer is that this entity has the same property *at different times*. Someone might be in Athens at an earlier time and at Crete at a later time. A difference in time need not lead to a change; someone can be in Athens at one time, and in Athens at a different – later or earlier – time. But it makes change possible – or, in a saying of the physicist Wheeler: "Time is nature's way of keeping everything from happening at once". Thus, time and change are closely connected to each other.

Changes can be of different sort, depending on the type of property at stake. There are, in particular

a) spatial properties, such as being in Crete and in Athens, being here and there, being under the blanket or on the blanket;

b) qualitative properties, such as being red and green, odd and even, immortal and mortal;

c) quantitative properties, such as getting a bit drunk or heavily drunk, driving seven miles or 99 miles, weighing one ton or nine tons. These properties are usually somehow derived, since they operate on qualitative or spatial changes and indicate differences in degree – either numerically or in a somewhat fuzzier way.

Accordingly, there a different types of changes – spatial, qualitative, quantitative. Any such change is a combination of times and properties. Motion, for example, is a change of spatial properties. Note that the property itself does not change, nor does the time change, although we often say this. What changes, is the assignment of properties to something, for example a person or an object, or perhaps to a full situation. If someone is alive at some time and dead at a later time, i.e., dies, then the property "be alive" does not change; but it so happens that this person does not have it any longer. Similarly, if someone grows from five feet to six feet, the properties "be five feet tall" and "be six feet tall" do not change; but the person has them at different times. And, of course, the earlier time is not all of a sudden a later time. In a word: neither times nor properties change; what changes is the assignment of properties to something over time.

The distinction between time and change seems an obvious one. But it has led, and still leads, to substantial confusion. In what follows, I will consider two examples that have played an important role in the discussion of time. The first confusion concerns the notion of *irreversibility*, that is, the "arrow of time" discussed above (section 2, point E). The laws of classical as well as of relativistic physics apply equally from earlier to later and from later to earlier: time is "reversible". Biological time – and similarly the time of our daily experience – is orientied: it runs from birth to death, never from death to birth; it is "irreversible". But this common way to state the difference is misleading. It is not time that is reversible or irreversible but the sequence of changes. Take the simple case of a glass which, once broken, cannot return to the state in which it was not broken (and thus really a glass, and not a mass of pieces): we can image this, even see it on a film that runs backwards; but it is never observed in reality. But suppose reality would indeed allow this to happen – then, there is still an earlier state, in which the glass is not broken, and a later state, at which it is not broken, interrupted by a state in-between, at which it is broken: there is a temporally ordered sequence "not-broken – broken – not broken", each of which is associated with a different time: there are three different time spans, two of which – the first and the last – are associated with the same qualitative properties. The glass does not "return in time" – it has the same properties again a later time. So, the difference between "reversible" and "irreversible" is only whether we can have the sequence of changes "unbroken to broken" as well as the sequence "broken to unbroken", or only the former. But each of these two sequences goes from earlier to later. Even if the order of changes can go in both directions – time cannot: there is no irreversable time.

The second confusion concerns the notion of a *cyclic time* (in contrast to a *linear time)*. From the Greek notion of the "Great year" – a very long period after which everything is destroyed by fire and then reborn – to Friedrich Nietzsche's "ewige Wiederkehr des Gleichen" ("eternal return of the same"), many share in the view that the world passes through cycles of creation, destruction and recreation (the classical treatment is Eliade 1954). In anthropology, it is often said that some cultures or some schools of thought do not have the western notion of linear time: there are time cycles (Wendorff 1980). In linguistics, Benjamin Lee Whorf became famous because of his claim that the Hopi have a completely different view of time than the one found in "Standard European Languages", a view which does not see time as a linear sequence but as a cycle (see the critical examination in Malotki 1983). But in all of these cases, this does not imply that the *time* is cyclic. It only means that the same *sequence of changes* is repeated and thus cyclic. The experience of such change cycles is very natural, on a short scale, as the sequence or day and night, as well as on a larger scale, as the re-appearance of certain stellar constellations. But this does not mean that the time itself comes and goes. We can count the repetitions. The seventh time, when the sun rises in the east, is not the same time as the twelfth time at which this happens. The twelfth time at which the world is re-created is not the fifteenth time at which it is re-created. What might be identical, are the properties which the world has at these different times.

Aristotle, to whom we owe the first systematic examination of time in general and of its relation to change in particular, states this very clearly in his famous definition of time: time is "a number of motion with respect to before and after" (Physics IV, 219 b 1–2). Aristotle's analysis of time is not easy to follow, and it has given rise to various interpretations (see, e.g., Coope 2005). But he not only makes a clear distinction between time and change; without such a distinction, it it would make no sense to say that something happens slowly or fast. He also characterises time as something that can be *counted*. If the entire world is reborn for the seventh time, then this seventh time is *later* that the sixth time at which it is reborn.

3.2. Time and its units

Is there one time, or are there many times? The common idea is that there is one time which can be subdivided into smaller units. These "smaller times" are called time spans, temporal intervals, subtimes, or just "times". We say that there is a time at which we met our first love, and a time at which we lost her or him, and each of these times it is a subinterval of the entire time; these subintervals themselves have subintervals: time is somehow nested. Several things can be said about these time spans:

1. They have a *duration*. We do not know whether the "entire time" has a duration. As we have seen above (section 2, point D) Newton equated absolute time with duration; but as soon as he talks about the measurement of time, only smaller time spans – for example the duration of some event – are at issue.

2. This duration can be measured. This is done by relating the time span, which is to be measured, to the duration of some other time spans; these are given by repeated occurrences of specific events (such as the rotation of the earth). We say that time can be measured. In a way, this is puzzling. When can it be measured? Clearly, time does not stand still during the measurement process – thus, the entity to be measured changes during this process. But there is no real puzzle: we do not really measure time, we measure the

duration of events, and we say that this duration is the time during which the event lasts. But one thing is the event, another thing is its time, just as there is a difference between a cup and the space which the cup occupies.

3. Time spans do not stand alone. They are related to each other according to an underlying earlier-later order which is unidirectional. But two time spans can also be simultaneous or overlap. In other words, time is a sort of structure whose units are time spans and whose structure is defined by temporal relations such as *succession, overlap, simultaneity*.

4. Each time span in turn consists of time spans. Does this go on forever – i.e., is there is "shortest time"? This is surely the case for human time experience. It is less clear whether nature has a minimal time span. Traditional as well as modern physicists assumed that *natura non fecit saltus* ("Nature does not make jumps", i.e., there is continuity). It was only in 1900 that Max Planck showed, quite reluctantly, that physicists are well advised to assume that there is a shortest time, whose duration is 5.4×10^{-43} seconds. We can, of course, *imagine* a shorter time, for example, 10^{-44} seconds; such a product or our mind is just meaningless for the laws of physics – and it would still leave us far away from a continuous time, which has no shortest interval.

5. Is there is a "last time span", i.e., does time have an end? And similarly, is there a "first time span", i.e., does time have a beginning? St. Augustine says no, Stephen Hawking says yes, Immanuel Kant says that both views lead to paradoxes. The reasonable person has no opinion on this issue.

3.4. Time and the observer

Neither physical time nor biological time, in the senses mentioned in section 2, know the distinction between past, present, and future – notions which everybody feels to be fundamental to human time. Einstein, in a conversation with Rudolf Carnap (around 1953), explicitly noted this fact: "Once Einstein said that the problem of the Now worried him seriously. He explained that the experience of the Now means something special for man, something essentially different from the past and the future, but that this important difference does not and cannot occur within physics. That this experience cannot be grasped by science seemed to him a matter of painful but inevitable resignation. I remarked that all that occurs objectively can be described in science; on the one hand the temporal sequence of events is described in physics; and, on the other hand, the peculiarities of man's experiences with respect to time, including his different attitude towards past, present, and future, can be described and (in principle) explained in psychology. But Einstein thought that these scientific descriptions cannot possibly satisfy our human needs; that there is something essential about the Now which is just outside the realm of science. " (Carnap 1963: 37f.). This distinction between past, present and future requires an observer; this observer cannot be an instrument which measures time, such as a clock. No chronometer, precise as it may be, distinguishes past from future. To this end, an observer is needed who identifies a time span as "being now". Human beings are able to do that. Maybe other animals are able to do it as well, although this question is not easy to answer.

But what is the "now"? In the long philosophical debate on this question, there has never been an answer on which the experts agree. Essentially, there are two different though interconnected problems. First, there is not just one "now" but infinitely many – the "now" right now, the "nows" that before that, and the "nows" that are ahead of us. In other words, time itself seems to be a series of nows. Acccordingly, there is a past and a future relative to each of these "nows". But what distinguisthes the "now" right now from all other "nows"? It must be a special property which somehow comes from the particular observer who experiences the – inner or outer – world as somehow "present", whereas earlier nows are somehow in memory, and later nows somehow in imagination. But on the other hand, the earlier "nows" are also defined in relation to the experiences of some observer, perhaps the same observer at some earlier time; so, the problem cannot be easily reduced to the difference between memory, experience and expectation in our soul, as St. Augustine does. It appears, therefore, that the distinction between "now" and "not-now" is not reducible to any other difference. The second classical problem results from the fact that the now does not seem to have an extension; otherwise, it would consist of several moments, some of which are earlier and hence past, and hence not now. But if the now has no extension, it does not exist, and hence, there is no presence; but if there is no presence, there is neither past nor future. Moreover, it is not possible that the entire time is built up from a series of nows, because if they have no extension, time cannot have an extension, either. These are the type of mind-boggling problems that were extensively discussed from Aristotle (Turetzky 1999: 22–25) to our days (see, for example, Dummett 2000).

In the second half of the 19th century, physiologists and psychologists set out investigate the notion of "present moment" with experimental methods. From film watching, everybody knows that when the number of pictures presented to our visual system exceeds about 20 per second, it cannot keep them apart and perceives them as a continous movement. So, there is a shortest time for (in this case visual) experience. But does this shortest time correspond to the "now" which underlies the distinction between past, present and future? When watching a film, or listenening to a tune, our intuition about what is on-going, rather than gone or only to come, seems much longer. So, there must be something in our brain which integrates shortest perceptual moments into a whole – a "perceptual present"; assumptions go that this perceptual present can last a few seconds.

Still a different issue is the "now" which underlies the linguistic expression of past, present and future. All languages in the world mark such a distinction by tense marking (*he is here vs he was here vs he will be here*) or by adverbials such as *yesterday, last year, very soon*. How is this now defined? Clearly, it cannot be the meaning of a word such as *now* (or its counterparts in other languages). These words refer to a time span with, as the case may be, considerable extension (*As a child, I was very religious, but now, I am not*). The word *now,* when uttered in a speech situation, refers to a time span which INCLUDES the moment of speech, rather to the moment of speech itself; the boundaries of this time span can vary. It seems to be this moment of speech which serves as an anchoring point, in relation to which present, past and future are defined. In fact, this picture is too simple again because the "moment of speech" is usually not a moment – surely not in the sense of the shortest time our brain can experience. We shall return to this problem in section 4. Two facts should be noted, however. First, this temporal anchoring to whatever the the "present moment" is usually considered to be fundamental to the expression of time in natural languages. Second, the temporal anchoring point may differ considerably from what in other disciplines is considered as "now".

4. The time concept of human languages

As we have seen in the preceding two sections, there are many notions of time, such as biological time, Newton's absolute and relative time, time as Kantian "Form der inneren Anschauung" and hence a necessary precondition of all cognition, subjective time, as influenced for example by drugs, and so on. These notions are interrelated in many ways, but they cannot be reduced to one concept: there are many. Is there a concept of time which underlies the expression of temporal relations in NATURAL LANGUAGES? Even this is doubtful. In most modern cultures, metrical calendar time plays an important role, so important that we often take it for self-evident. Our life is largely organised around (or rather along) this time, and hence, there are many expressions which refer to it – like *in the year of 2007, two hours and thirty five minutes after noon on May 8, 1998*, and so on. But many cultures do not have such a concept of metrical time, nor the notion of one major event in collective history to which everything can be temporally related. Even in Western culture, the full elaboration of this system is fairly recent. The mere fact that people talk of "hours", "days", "years" and "the birth of Christ" does not mean that they have a concept of metrical time, with the birth of the saviour, or some other important event, as point zero. Until a few centuries ago, the concept of "hour", for example, just meant "twelfth part of the day", and if the day was short, like in winter, the hour was short, as well. A "day" is simply the time, when there is light, or the time from when people get up until they go to bed again, no matter how "long" this may be in terms of a mechanical or electronical clock.

Therefore, it seems reasonable to distinguish between various layers of time structure that are used in the encoding of time. There is something like a "basic time structure" on which the expression of temporal relations in natural languages is based. This basic time structure must cover basic relations between time spans, such as succession and simultaneity, but also the notion of a basic vantage point – the "now" of an observer. More differentiated structures, like calendaric metrical time, may be added, as cultures develop. It seems likely, although this is an empirical question, that such additional structuring is only expressed by more or less complex lexical expressions, whereas the basic time structure is most often expressed by grammatical categories and by simple adverbs.[7]

4.1. The "basic time structure"

4.1.1. The ingredients

What, then, is this "basic time structure"? This is not easy to say, because at most 5% of the world's languages are sufficiently well described; for all others, our information is very superficial and often based on bold comparisons with familiar languages such as Greek, Latin or English. Hence, we might simply miss important temporal notions encoded in same or even many languages. But such is the state of our knowledge. An inspection of

---

[7] The following discussion essentially follows Klein (1984), Chapter 4.

those languages for which our information is more profound shows that the following six characteristics are indispensable:

A. Segmentability: Time, whatever it is, can be divided into smaller segments – "time spans" or "temporal intervals".

As was discussed in section 3.3, there is a perennial debate among philosophers and physicists on whether this division can be infinitely repeated or whether there is some minimal "time quantum". I do not believe that the mind of the common language user has a standing on this issue, and in fact, I would not know of any criterion to decide whether we need infinite segmentability, if we want to describe the linguistic expression of temporal relations. Let us now turn to these relations between time spans.

B. Inclusion: If s_1 and s_2 are time spans, then s_1 may be included in s_2; this inclusion may be full or partial; in the latter case, we may speak of "overlapping".

C. Succession: If s_1 and s_2 are time spans, which are not (fully or partly) included in each other, then either s_1 precedes s_2 or s_2 precedes s_1.

It is usually said that time is linearly ordered. The way in which we have characterised succession here is somewhat weaker: there is a partial order on time spans: time spans can overlap. Again, it is an open question whether this partial order is based on some full order on "time points", which make up the time spans. We normally assume that there is some temporal progression within a time span, and a strict order on time points allows us to reconstruct this intuition in a straightforward way.

These three features allow a clear definition of the "earlier-later" asymmetry between time spans as well as simultaneity. Simultaneity can be full (two time spans completely coincide) or partial, if they partly overlap.

D. Duration: Time spans may be long or short in duration.

Duration, as regularly expressed in natural language, is not another name for time, as in Newton's definition. It is a property of time spans. It is typically indicated by adverbials, such as for *two days, rapidly, quite a while*. They do not necessarily describe objectively measured time. If we say *It took Shin quite a while to...*, then we may refer to very different "objective durations", depending on whether we talk about drinking a coffee or finding a spouse.

E. Origo: There is a distinguished time span, which we may call "the time of present experience". Everything before that is accessible to us only by memory, everything later only by expectation.

This origo is the dividing point between past, present and future. As was discussed in sections 2. (points H. and I.), and 3.3, such an origo is not part of all time concepts; it plays no role in physical time or in biological time. But it plays an eminent role in the linguistic encoding of temporal relations. The best-known case is the grammatical category of tense; in its classical understanding, tense situates some event in relation to the "deictic origo", which is given by the moment of speech – the linguistic variant of the time of present experience. But there are also many adverbials which are anchored at the deictic origo, for example *today*, *three days ago* or, of course, the word *now* itself; thus, *today* means "the day which includes the deictic origo". Remember, however, that the meaning of the word *now* is not to be equated with the deictic origo – it refers to a time span which contains the deictic origo, but can be much longer (cf. section 3.3). We can say, for example *Now, the average temperature is much colder than in the pleistocene.*

F. Proximity: If s_1 and s_2 are time spans, then s_1 may be near to, or far from, s_2.

This feature is much less discussed in the tradition than linear order, duration, or the existence of a "now". But it is regularly encoded in natural languages. Proximity and non-proximity in this (non-metrical) sense is exemplified, for example, by expressions like *soon* or *just*; it also sometimes shows up in tense distinctions, like "near future" vs "far future". Note that this concept of "temporal distance" or "remoteness" does not presuppose a concept of metrical time. Quite the opposite, it is not easy to capture the idea of proximity in this sense by metrical distance: *soon* can mean "in ten minutes", like *the meal will be served soon*; but it can also mean "ten months", like in *they soon got divorced again*.

G. Lack of quality: Time spans have no qualitative properties; they are neither green nor sweet, and they have no wheels and no spines. They are contained in each other or just after each other or or more or less close to each other, and they are long or short.

In the tradition, this feature shows up in the discussion about how time and change are related to each other (see section 3.1 above). The latter normally relates to changes in qualitative properties or position, the former to the "pure structure", in relation to which such changes are perceived, imagined, or expressed. When we talk about time, then typically, some descriptive properties are *associated* with certain time spans – for example, we may talk about the time at which some event took place, or some state obtained. But we must carefully distinguish between an event or a state, and the time at which these take place or obtain.

4.1.2. A more precise definition

The usual way to give a precise definition to temporal relations is to interpret time spans as closed (sometimes as open) intervals of the real numbers; the "smaller than"-relation between real numbers is then used in the obvious way to define a partial order on the intervals (if $s = [r_i, r_j]$ and $t = [r_k, r_l]$ are closed intervals, then s is BEFORE t iff $r_j < r_k$). This procedure, whilst straightforward and elegant, is not sufficient, however. It provides us both with too much and too little structure. Under the assumptions made in 4.2, the Basic Time Structure does not include the notion of a metrical distance between time spans; the definition just sketched does not, either; but the underlying relation between "time moments", identified with the real numbers, does. It also makes the assumption that time is dense, i.e., that there is no smallest time span, an assumption which may be too strong. But these problems are perhaps not really harmful. It is much more problematic that some crucial intuitions are not captured, in particular the features "origo", "proximity", and "duration". Hence, we need "more" structure.

(a) The most straightforward way to account for the notion of "origo" is to identify it with the moment of speech; in fact, such a "deictic origo" is found in all human languages we know.

(b) It is less clear how one should capture our intuitive notion of (temporally) nearness. One might think to use the natural topology on the real numbers: the neighbourhoods of any real number r are exactly those open intervals to which r belongs. But this gives us by far too much: it gives us all environments, rather than the one which marks the borderline between "close" and "far". Our intuitive notions tell us that each time span has a "REGION" around itself, whose borders vary with context. The time of drinking a coffee is usually shorter than the time of finding a proof, and so are the "regions" around these two time spans. Temporal relations between two time spans s and t do not only differ according to whether s precedes t, follows t, or is (partly or fully) contained in t, but also according to whether it is "in the region of t". This region may be very wide, if t itself is "long"; but it may also be short. It may also happen that the region is lexically or grammatically specified.

(c) There is no such straightforward solution for the related problem of duration. The fuzziness of durational notions like *for a while, shortly, very much later* cannot be accounted for by metrical time, on the one hand, nor by introducing simply a "region" around time spans. In some cases, one can relate the relative duration of a time span, for example the time which some event takes, to the average time of similar events. For example, in *She rapidly drank a beer*, the time of this beer-drinking is related to the average time of beer-drinking and found to be shorter than this average time. But there are cases in which this does not work, like in *He slept for a while* as compared to *He slept for quite a while*. It is no surprise, therefore, that the meaning of these expressions is hardly ever precisely described.

The components of the Basic Time Structure are thus:

- an infinite set of time spans (leaving aside whether these are infinitely divisible)
- an order relation on time spans (BEFORE)
- a topological relation IN between time spans
- for each time span t, a distinguished time span which includes t – the REGION of t
- a distinguished time span, the ORIGO.

We may now define the Basic Time Structure as follows:

(1) The Basic Time Structure (BTS) is a structure [$\mathbb{R}$, $\{t_i\}$, $\{R_i\}$, BEFORE, IN, 0], where
 - $\mathbb{R}$ are the real numbers, with the usual order relation <
 - $\{t_i\}$ is the set of closed intervals of $\mathbb{R}$, the "time spans";
   $\{R_i\}$ is a subset of $\{t_i\}$, such that for each t_i, there is exactly one R_i which properly includes t_i (R_i is the REGION of t_i);
 - BEFORE is a partial order on $\{t_i\}$, such that: If $s = [r_i, r_j]$ and $t = [r_k, r_l]$ are in $\{t_i\}$, then s BEFORE t iff $r_j < r_k$;
 - IN is a relation on $\{t_i\}$, such that s IN t iff they have at least one element in common
 - 0 is a distinguished element of $\{t_i\}$, the ORIGO.

The Basic Time Structure is a sort of scaffold which allows us to define various types of temporal relations such as BEFORE, AFTER, IN. These relations obtain between two time spans, which I will call temporal relata. In *John left yesterday*, for example, one of the relata is the time of John's leaving, the other relatum is the time at which the utterance is made, and the relation is BEFORE. Other, much more complex constellations are possible. In what follows, I will first illustrate some characteristic relations[8] and then discuss the various types of temporal relata.

4.2. Temporal relations

Temporal relations obtain between two time spans: a first time span, which I will call THEME, and some other time span, which I will call RELATUM. The what follows, theme is marked by ------, the relatum is marked by +++++, and the region around the relatum by ( ); the linear order is represented by left-right arrangement:

a. BEFORE, i.e., the theme precedes the relatum properly:
      ------   +++++

b. LONG BEFORE, i.e., the theme precedes the region of the relatum:
      ------ (+++++)

c. SHORTLY BEFORE, i.e., the theme precedes the relatum, but it is in the region of the relatum:
      (------ +++++)

d. JUST BEFORE, i.e., as SHORTLY BEFORE, but the theme abuts the relatum:
      ------+++++

In this case, the theme is automatically in the region of the relatum – more precisely, the final part of the theme; in principle, it is not excluded that the theme begins long before the relatum.

---

[8] In all of these cases, the Basic Time Structure allows us to give precise formal definitions. For present purposes, however, it will be more useful to use diagrams that illustrate the various relations.

e. PARTLY BEFORE, i.e., a the first part of the theme precedes and the second part of the theme is IN the relatum (the region is irrelevant):
   ---+-+-+-++

f. INCL, i.e., the theme is fully included in the relatum:
   ++-+-+-+-+-+++

g. AFTER, i.e., the relatum precedes the theme:
   +++++------

Other relations, such as JUST AFTER, SHORTLY AFTER, LONG AFTER can be defined analoguously. Note that the relation IN has been split here into PARTLY BEFORE, INCL and PARTLY AFTER; if we want such a notion, it can be defined by the usual Boolean operations.

4.3. Temporal relata

When a temporal relation is expressed in some communicative situation, the two relata normally have a different functional status. One of them, for example the time of some event, is somehow "situated" in time; this is done by relating it to some other time span which is supposed to be given in the communicative situation and then functions as a kind of anchoring point. I shall call the former, the theme, and the latter, simply the relatum, respectively. The familiar grammatical category of tense exemplifies this functional asymmetry very well. It indicates, at least in its classical understanding, that some event is in the past, present or future – that is, it precedes, includes, or follows the moment of speech. Thus, in *John left*, when uttered in a particular communicative situation, the time of John's leaving is the theme, and the moment of speech is the relatum. Basically, there are three ways in which such a relatum can be given:

– deictic, that is, it can be derived from the speech situation;
– anaphoric, that is, it is mentioned in the preceding context;
– calendaric, that is, it given by some important event in cultural history

Calendaric relata is of lesser interest here; they only differ in which historical event from the shared knowledge of the interlocutors is chosen as an anchoring point – the foundation of Rome, birth of Christ, the Hedjra, the beginning of a dynasty, etc. There is no language in which tense is linked to a calendaric origin. But many languages have a rich system of adverbials with such an anchoring point.

4.3.1. Deictic relatum

The Basic Time Structure, as defined above, includes a distinguished time span, called there the origo, which plays a special role in the expression of time. What is the origo in a given communication? Typically, it is identified with the "moment of speech" or, as is often said, the "time of utterance". The latter expression is preferable, since the "moment of speech" is ususally not just a moment. Expressions which use this time of utterance as a relatum are usually called "deictic". The verbal category of tense, which is deeply rooted in the grammatical system of many – though not of all – languages is deictic: *He was singing, he is singing, he will be singing* place the time of some event, before, around, or after the time of utterance. But deictic relata also underly many adverbials. Thus, *three years ago* (in contrast to *three years before*) means "at a time which is three years before the time of utterance", and *yesterday* means "at the day which precedes the day which includes the time of utterance".

The deictic relatum is fundamental to many temporal expressions. But it also raises a number of problems, three of which I will briefly mention here. First, how long is the "time of utterance"? Does it include the whole interval during which an utterance is spelled out, is it only a part of the latter, or is it even longer? Sometimes, a shorter relatum is needed, for example when someone says:

(2)   From *now*, it is precisely four seconds until *now*.

We also have the opposite problem, i.e., cases in which the "time of utterance" seems to go beyond the boundaries of a single sentence. Does a longer text, say a lecture or even a novel, have a single time of utterance or a different one for each single utterance of which it consists? In a sense, a coherent sequence of utterances – a text, be it written or spoken – is a unit, and it should have a single relatum. But then, it would be strange to assume that this relatum is, for example, the time at which the whole text was produced: What is then the utterance time of the Bible, or the first book of Moses? In these cases, the characterisation of the deictic relatum as "time of utterance" is clearly insufficient.

Linguistic systems most often evolve in spoken communication, in which speaker and listener are equally present. Then, time of speaking and time of hearing collapse, and hence, there is no need to distinguish between the speaker's and the listener's origo. But in other communicative situations, there are clashes, for example in written language (or even in spoken language, when it is stored in some way). In this case, it is regularly the speaker's origo which counts.

The third problem with the notion "time of utterance" concerns possible shifts – i.e., cases in which it is not the origo (the time of present experience) which counts but rather some other time interval. Two such cases are usually mentioned in the literature. The first is exemplified by "vivid narration", like in the historical present, in which the speaker treats past events as if they were happening now. Somehow, the time of utterance is replaced by the time of actual experience; it is the latter which serves as relatum. The other kind of shift is introduced by verbs of saying and thinking, as in these examples:

(3) I thought: Now, I must change my life.

(4) Yesterday, my friend said: Shouldn't we go to Berlin tomorrow?

In these cases, it is not the origo of the real speaker which counts but the origo of the person whose thinking or speaking is being talked about.

4.3.2. *Anaphoric relatum*

Anaphoric relata are time spans which are given somewhere in the linguistic context. Their role for tense is disputed. In the literature, a distinction is often made between "absolute" and "relative" tenses; the former are purely deictic, whereas the latter also involve an anaphoric relatum. Some text types, for example narratives, are based on an chain of such anaphoric relata. As with all types of anaphora, there are three subcases:

1. The anaphoric relatum is within the same clause (intraclausal anaphora):

In (5), the initial adverbial introduces a time span, to which another time span in the same utterance is related:

(5) At six o'clock sharp, he switched the light off.

2. The anaphoric relation may go from one clause to another, whilst still being in the same sentence (interclausal anaphora):

(6) When the phone rang, he switched the light off.

In cases of this type, it is often said that "two events" are temporally related to each other. But note that the entire *when*-clause only serves to define a time span, which functions as a relatum. In principle, this is not different from the anaphoric relatum in (5), which is simply specified by a clock-time adverbial.

3. The anaphoric relatum may have been introduced in a preceding utterance:

This type of anaphoric temporal linkage is most import for text organisation. It is exemplified by well-known discourse principles such as "the principle of chronological order" which states that, unless marked otherwise, the time span of some situation described is after the time span of the situation mentioned in the preceding utterance.

A time span that functions as an anaphoric relatum for some subsequent time span may in itself be based on a deictic relatum. Compare the following two intraclausal anaphoric relata:

(8) Three weeks ago, he didn't have a penny.

(9) Three weeks before that, he didn't have a penny.

In both cases, the initial adverbial introduces a time span, say t_6 and t_7, respectively. In the first case, this time span t_6 is three months before the time of utterance, in the second, t_7 is three months before some other contextually given event. Hence, the time span is deictically given in (8) and anaphorically in (9). But in both cases, the time span functions as an anaphorical relatum of the subsequent time span – the time at which he had no penny. The fact that something is an anaphoric relatum of something else does not preclude that it is in itself deictically introduced. On the other hand, we may often get "anaphoric chains"; an anaphoric relatum is temporally related to a preceding one, which in turn is related to another one, and so on, and so forth.

5. Concluding remarks

The ability to talk about time is a fundamental trait of human communication, and all languages we know of have developed means to express time. But in sections 2 and 3, we have seen that time is not a uniform phenomenon. There are numerous concepts of time; they share some features, but they are also divergent in many respects. Which of these concepts underlies the expression of time in human languages? There is no straightforward answer, for at least two reasons. First, we are not well informed about most languages of the world. Second, those languages we know seem to differ in what they encode and how they do it. One way to approach both problems is to start with a relatively simple "basic time structure", which covers the core notions expressed in same of the better-studied languages. In section 4, such a basic concept is defined. As need arises, it can be refined; it can also be simplified, if the language to be described does not use all of features of this structure. But may serve well as a point of departure.

Acknowledgement

I wish to thank Leah Roberts who corrected my English.

References

Aristotle (many editions)
   350 BC   *Physics*. Available at http://classics.mit.edu/Aristotle/physics.html.
Bruton, Eric
   1993   *The History of Clocks and Watches.* London: Black Cat.
Butterfield, Jeremy (ed.)
   1999   *The arguments of time.* Oxford: Oxford University Press.
Carnap, Rudolf
   1963   *Intellectual Autobiography.* In *The Philosophy of Rudolf Carnap*, P. A. Schilpp (ed.), 1–84. La Salle, IL: Cambridge University Press.
Coope, Ursula
   2005   *Time for Aristotle.* Oxford: Oxford University Press.
Dietrich, Rainer
   2007   *Psycholinguistik.* Stuttgart: Metzler.
Dohrn-van Rossum, Gerhard
   1996   *History of the Hour: Clocks and Modern Temporal Orders.* Chicago: University of Chicago Press.
Dummett, Michael
   2000   Is Time a Continuum of Instants? *Philosophy* 75: 497–515.
Eliade, Mircea
   1954   *Cosmos and History: The Myth of the Eternal Return.* Princeton, NJ: Princeton University Press.

Flasch, Kurt
    2004   *Was ist Zeit?. Augustinus von Hippo, das XI. Buch der Confessiones.* Frankfurt a. M.: Klostermann.

Foster, Russell G. and Leon Kreitzman
    2004   *Rhythms of life: The biological clocks that control the daily lives of every living thing.* London: Profile Books.

Fraser, Julius T.
    1987   *Time – the familiar stranger.* Amherst: University of Massachusetts Press.

Fraser, Julius T. (ed.)
    1968   *The voices of time.* London: Penguin

Fraser, Julius T., Francis C. Haber and Nathaniel M. Lawrence (eds.)
    1986   *Time, Science, and Society in China and the West.* Amherst: University of Massachusetts Press.

Heidegger, Martin
    1927   *Sein und Zeit.* Jena: Niemeyer.

Herder, Johann Gottfried
    1784   *Ideen zu einer Philosophie der Geschichte der Menschheit.* Riga: Hartknoch.

Horwich, Paul
    1987   *Asymmetries in Time.* Cambridge, MA: MIT Press.

Janda, Richard D. and Brian D. Joseph (eds.)
    2004   *The Handbook of Historical Linguistics.* Oxford: Blackwell.

Jourdain, Robert
    1997   *Music, the Brain, and Ecstasy: How Music Captures Our Imagination.* New York: William Morrow.

Kant, Immanuel
    1781   *Kritik der reinen Vernunft.* Riga: Hartknoch.

Klein, Wolfgang
    1994   *Time in Language.* London: Routledge.

Le Poidevin, Robin and Murray McBeath (eds.)
    1993   *The Philosophy of Time.* Oxford: Oxford University Press.

Libet, Benjamin
    2004   *Mind Time. The Temporal Factor in Consciousnes.* Cambridge, MA: Harvard University Press.

Lucas, John R.
    1999   A century of time. In Butterfield 1999: 1–20.

Malotki, Ekkehart
    1983   *Hopi Time.* Berlin/New York: Mouton de Gruyter.

McTaggert, John E. M.
    1908   The Unreality of Time. *Mind* 17: 457–474. (Reprinted in Le Poidevin and McBearth 1993).

Nakayama, Hajime
  1968   Time in Indian and Japanese thought. In Fraser 1968: 77–91.
Newton, Isaac
  1972   *Isaac Newton's Philosophiae naturalis principia mathematica.* The 3rd edition, with variant readings, assembled and edited by Alexandre Koyré and Bernard Cohen. Cambridge, MA: Harvard University Press.
Paul, Hermann
  1882   *Prinzipien der Sprachgeschichte.* 2nd Edition. Jena: Niemeyer.
Piaget, Jean
  1923   *Le développement de la notion de temps chez l'enfant.* Paris: Presses Universitaires de France.
Poeppel, Ernst
  1988   *Mindworks: Time and Conscious Experience.* Boston: Harcourt.
Prigogine, Ilya, and Stengers, Isabelle
  1993   *Time, Chaos and the Quantum: Towards the Resolution of the Time Paradox.* New York: Harmony Books.
Reichenbach, Hans
  1958   *The Direction of Time.* Berkeley, CA: University of California Press.
Richards, Edward G.
  1998   *Mapping Time, the calendar and its history.* Oxford: Oxford University Press.
Savitt, Steven, ed.
  1995   *Time's Arrows Today: Recent Physical and Philosophical Work on the Direction of Time.* Cambridge: Cambridge University Press.
Turetzky, Philip
  1998   *Time.* London: Routledge.
Whitrow, Geoffrey J.
  1980   *The natural philosophy of time.* Oxford: Clarendon Press.
Wendorff, Rudolf
  1980   *Zeit und Kultur. Geschichte des Zeitbewußtseins in Europa.* Wiesbaden: Westdeutscher Verlag.
Westfall, Richard Samual
  1983   *Never at rest. A biography of Isaac Newton.* Cambridge: Cambridge University Press.

How time is encoded

Wolfgang Klein

1. Introduction

The experience of time is fundamental to human cognition and action. Therefore, all languages we know of have developed a rich repertoire of means to encode time. In many languages, the expression of time is even close to mandatory, since it is structurally connnected to the finite verb:

(1)   a. Eva was cheerful.
      b. Eva is cheerful.
      c. Eva will be cheerful.

Each of these sentences positions a certain situation, Eva's being cheerful, before, around or after the moment of speech. We must clearly distinguish here between

- the **situation itself**; in (1), it is a sort of state; it could also be a short event, such as the one described by *Eva closed the door* or by a slow process such as *Eva grew older*. Following Comrie (1976), I shall use the word "situation" as an overarching term for all sorts of events, states, processes, actions, etc.
- the **description of the situation**; this description is realised by the non-finite parts of the sentence; I shall mark such a description as [...]. In (1), this description is [Eva be cheerful] in all three cases; this does not mean, of course, that the situation itself is the same, nor does it mean that the situation could not be described in some other way.
- the **marking of how the situation is positioned in time**; in (1), this is done by modifications of the verb, here by the choice between *was, is, will be*.

In English, the description of the situation and its positioning in time can normally not be separated. There is no non-finite utterance such as (1d):

(1)   d. Eva be cheerful.

Other languages, such as Chinese, do not force their speakers to mark time. This does not mean, of course, that they cannot locate situations in time. They just use other means, in particular adverbials such as *in the past* or *soon*.

Essentially, there are six types of devices that are regularly used to encode time in language. These are:

1. *Tense*. Tense is a grammatical category of the verb; in its traditional understanding, it serves to locate the situation in relation to the "now" of the speech act. Thus, the difference between *was, is*, and *will be* in (1) reflects different tenses.

2. *Aspect*. Aspect is also a grammatical category of the verb; in its traditional understanding, it serves to "present" a situation from a particular viewpoint, for example as on-going or as completed. Thus, *Eva was closing the door* presents a situation described by [Eva close the door] as on-going, whereas *Eva closed the door* presents a situation described by [Eva close the door] as completed; in both cases, the tense marking positions the situation before the moment of speech.

In principle, tense and aspect should be independent from each other, i.e., the same aspectual contrast could be found in all tenses. In English, this is largely the case (there are a few exceptions). In most languages, however, tense and aspect are combined to a simpler inflectional system – which is mostly called "tense system", rather than "tense-aspect system". In Russian, for example, the pure aspectual contrast between imperfective aspect and perfective aspect only applies to past tense forms (see section 3.3 below).

3. *Aktionsart* ("event types", "lexical aspect"). Aktionsart is traditionally considered to be a subdivision of verb types according to the temporal properties of the situations which they describe.[1] Thus, *to sleep* is used to describe a "state", whereas *to close* is used to describe an "event". The term is also used for more complex verbal expressions, such as *to fall asleep, to sleep for an hour* or *to close three windows*.

4. *Temporal adverbials*. These are by far the richest class of temporal expressions, and in contrast to tense and aspect, they are found in all languages. Temporal adverbials can be simple (*now, soon, often*), morphologi-

---

[1] Note that verbs never **refer** to situations, as is often said (which situation does *to sleep* refer to?). They are used to describe certain properties of situations.

cally compound (*today, rapidly, afterwards*) or syntactically compound (*after the war, long ago, when the saints go marching in*). Functionally, they can describe very different temporal features, such as position on the time line (*now, yesterday, next year*), duration (*for two hours*), frequency (*rarely*), and many others whose precise role is not easy to determine (*still, again*).

5. *Temporal particles*. They are somewhere between temporal adverbials and suffixes or prefixes; well-known examples are the Chinese particles *le, zhe* and *guo* which can follow the bare verb stem and which express something like aspect.

6. *Discourse principles*. Very often, temporal relations are not expressed by specific words or constructions but by the way in which sentences are organised into larger stretches of discourse. In ancient rhetoric, for example, there was a principle called "hysteron proteron (i.e., later-earlier)", which, in a nutshell, stated that in the default case, events in a story should be told in the order in which they occur.

There is an extensive research on these devices, their form, their function, and the way in which they interact in a sentence and in a larger piece of text (the best historical survey is still Binnick 1991). But it is perhaps fair to say that the agenda has not been closed on any of them. There are impressive findings, but there are also many gaps and insufficiencies. Overall, the investigation of how time is encoded in natural languages suffers from three substantial shortcomings:

A. It is strongly biased towards certain devices. From Aristotle to present times, there is a steady stream of research on tense and on Aktionsart; in fact, the way in which we think about the expression of time is deeply shaped by what the Greek philosophers thought about it, and thus, by the structure of Greek. In more recent times, this has been matched by studies on aspect. There is much less work on temporal adverbials, particles, and discourse principles. This is somewhat perplexing, because in contrast to the verbal categories tense and aspect, temporal adverbials are not only found in all languages but they also allow a much more differentiated expression of time than any other device. In fact, one wonders whether tense and aspect are not completely superfluous in view of what temporal adverbials allow us to do.

B. It is strongly biased towards certain languages. Most work deals with a few Indoeuropean languages, such as Greek, Latin, English, German or

Russian. Of at least 90% the world's languages, we have only very vague ideas on how they express time, often based on very superficial descriptions by missionaries who tried to find "analogues" to tense and aspect in these languages. Thus, notions such as "imperfective aspect" or "past tense" are somehow transferred to Arrente or Kpelle, although neither form nor meaning are necessarily the same as in English or Russian.

C. It is strongly biased towards certain text types. Most work by far on the expression of time deals with singular events in reality. Other text types, for example instructions, descriptions, laws – if dealt with at all – are analysed against this background. This is often problematic. Tense, the most important temporal category, is supposed to relate the "situation" to the moment of speech. But what is the moment of speech in a novel, a cake recipe, or a law – text types which are surely not exotic?

So, the state of the art on how languages encode time yields a very unbalanced and incomplete picture. A second, no less serious problem may already have become clear above in the informal characterisations of the six types of devices. There is an initial understanding of notions such as tense, aspect, or Aktionsart, shared by most linguists and grammar makers. But on closer inspection, it rapidly turns out that each of these notions is loaded with problems that range from terminological confusion to fundamental unclarities of definition. This will become clear as we now turn to the six devices in more detail. In what follows, the focus will be on the first three devices – tense, aspect, and Aktionsart. As has already been mentioned, this does not do justice to what really happens in human languages – but it mirrors our knowledge about it.

2. Tense

2.1. The canonical view[2]

In its received understanding, tense is a deictic-relational category of the verb: it indicates a temporal relation between the situation described by the sentence and some deictically given time span; this time span is usually the

---

[2] A very clear treatment, including a discussion of more recent developments, is Comrie (1985). Dahl (1985) and Bybee, Perkins and Pagliuca (1994) are very useful surveys.

moment at which the sentence is uttered – the moment of speech, the utterance time, or the "now". In what follows, I shall mostly speak of "time of utterance" (abbreviated TU) for this deictic anchor. Typically, three temporal relations are distinguished, and hence, it is often assumed that there are three basic tenses:

Past: The time of the situation precedes the utterance time.
Present: The time of the situation is more or less simultaneous to the utterance time.
Future: The time of the situation follows the utterance time.

This idea goes back to the Greek philosophers. With some refinements, it is still found in most descriptive grammars, and also in many treatises on tense. The word tense itself comes from Latin *tempus* ("time"). In some languages, such as French or Italian, one and the same word is used for time and tense, and in many other languages, the terms "past, present, future" refer equally to the **grammatical tenses** and to the **notions** of past, present, and future. This common origin easily invites the idea that one cannot properly express time without tense. This is, of course, a weird idea. After all, a language which does not inflect its verbs for tense can easily add an adverbial such as *in the past, now,* or *in the future,* let alone more differentiated characterisations such *as tomorrow, at this very moment,* or *in seven years from now*. So, tense is not only to be separated from time – it is not even a particularly important for the expression of time. Many languages do not have it at all, and in those languages which do have it, it is largely redundant. Still, it attracted, and probably still attracts, most attention in linguistic research on time. This research has pleased us with many insights. But there are also many problems left, to which I will now turn.

To begin with, there are many terminological confusions. The word "tense" is used in at least three ways: it can refer to

– the grammatical notion tense (as in *Tense is an inflectional category of the verb*),
– a particular form, such as in *The English past tense is marked by* ... In this sense, "tense" often links a particular form to a bundle of functions, such as "pastness" and "perfective aspect"
– a particular function, as *This form expresses future tense*. In Russian, for example, the "present tense form" of the perfective aspect has a "future tense meaning".

These terminological confusions could easily be avoided by a more careful choice or wording (for example by distinguishing between "tense form" and "tense meaning", as need arises). In actual fact, however, they often lead to confusions, for example when people speak about the "past tense" in sentences such as *If he arrived before eight, the party could begin in due time*. Another example of this terminological nuisance are the fruitless discussions on "how many tenses" a particular languages has. In German, proposals range from 1 to 12 (see Thieroff 1992 for a discussion), depending on what exactly is meant by "a tense".

More serious, however, are four fundamental problems with the classical notion of tense: there are more than three "tenses", the classical definition is often wrong, the temporal anchor need not be the moment of speech, and there are many "non-canonical" uses. We will now look at these four problems.

2.1. There are more "tenses" than past, present and future

Under its simplest definition, tense marks past, present, and future: the situation is before TU, it is at TU, it is after TU. Now, even in Greek, for which this idea was originally invented, there are more than three "tenses" (in the sense of tense forms). So, how can we capture the meaning of these forms, if is not just the bare distinction between past, present and future?

Over the centuries, several ways to overcome this problems have been developed, in particular the following three:

A. *More differentiated temporal relations*. There is not just a relation "before", but relations such as "long before", "shortly before" or "before but on the same day" vs "before but not on the same day". There are indeed some languages which seem to have such refined temporal notions (Dahl 1985).

B. *Additional time spans*. Past tense, present tense and future tense in the classical sense express a relation between two time spans – the time of the situation and the time of the utterance. It could be that there is a "third time", which is, for example, between these two. Implicitly, this idea is found already in the traditional understanding of tense forms such as the plusquamperfect of the second future of Latin. It was more systematically elaborated in the course of the 19th century. Hermann Paul (1882: 273f.), for instance, stated that there is a special "vantage point", from which the situation is seen. This vantage point may precede or follow the time of utterance, and it may precede or follow the time of the situation, thus giving rise to more compound relations. Half a century later, this idea was taken up by the philosopher Hans Reichenbach (Reichenbach 1947), who dubbed the three "points" S, E, and R, respectively. S corresponds to the time of utterance, E to the time of the situation, and R is the "point of reference", which corresponds to Paul's vantage point. He applied them to the English tense system. Reichenbach's labels were taken up by many other authors, although mostly with the understanding that they do not refer to "time points" but to "time intervals". This type of analysis has become very popular, although it suffers a bit from the fact that Reichenbach never bothered to explain what he meant by "point of reference" (see, e.g., Hamann 1987, Klein 1992).

C. *Combination of temporal relation and other temporal function.* A "tense form" may not just express how the situation is related to the TU but also invoke a certain perspective on the situation. This may apply to the two English "past tenses" *He was sleeping* and *He slept.*, or to the three French "past tenses": *Il parla, il parlait, il a parlé.*

D. *Non-temporal functions of "tense forms".* In German, the so-called future tense is ambigouous between a temporal reading and a modal reading. Thus, *Hans wird schlafen* "Hans will sleep" can mean: "He will sleep at some time after now", but it can also mean: "It is likely that he is sleeping now". In English, the "past tense" may express pastness, but also irreality, as in *If they were here, ...*

All of these possibilities are indeed observed in the tense systems of various languages. This fact does not speak against the idea that tense expresses a temporal relationship between a situation and the time of utterance; but it shows that the picture is much more complex than the classical notion of past, present, and future would suggest.

2.2. Relation between what?

Let us now ignore these complications and return to a very simply case, such as (1a), *Eva was cheerful.* Under the classical understanding of tense, the past form of the verb *was* indicates that a situation described by [Eva be cheerful] is before TU, whereas *Eva is cheerful* says that a situation described by [Eva be cheerful] it more or less simultaneous with TU. It is easy to see that this cannot be correct. If *Eva was cheerful* is true, then this does not exclude at all that she is still cheerful at TU. If *Eva's cat was dead* is true (let's hope it is not!), we can be quite certain that the cat is still dead at the moment of speech. So, if the time of the situation is that time at which the situation obtains or happens, then the classical definition of tense is bluntly false in very elementary sentences.

When someone asserts *Eva's cat was dead*, then he asserts something about some time span in the past – the time talked about, the assertion time, or the **topic time**, as I shall say. This time can, but need not, be the time at which the situation obtains or happens. In *Eva's cat was dead*, the topic time is most likely a subinterval of the time of the situation, that is, the time at which the cat is dead. So, even in elementary tense forms, three time spans come into play:

- the time of utterance; this is the time at which the utterance is expressed
- the topic time; this is the time about which something is asserted (or asked)
- the time of the situation; this is the time at which the situation obtains or occurs

What tense does, is to express a relation between the time of utterance and the topic time – the time about which the speaker wants to say (for example to assert) something. This topic time in turn is temporally related to the time of the situation: it can be contained in it, it can contain it, it may follow or precede it, it may also be fully simultaneous to it. In this latter case, the classical notion of tense is correct: it marks a relation between between the time of the situation and the time of utterance. In all other cases, there is only a "mediated relation" between these two times.

2.3. The temporal anchor need not be the time of utterance

Under the classical understanding, the temporal anchor is deictically given: it is the "now", the 'moment of speech', or the 'time of utterance' TU. This idea, firmly established as it is, reflects the various biases which characterise the research tradition on tense (see the discussion at the end of section 1). As soon as we go beyond the conventional diet of examples, it faces at least four problems.

First, it often does not work for subordinate clauses. In many languages with tense marking, the present tense of sentences like (2) can refer to John's thinking time – which is in the past – or to the time at which the sentence itself is uttered:

(2)   John thought that Mary is in the kitchen.

Under the "matrix clause anchoring", the situation of the subordinate clause is in the past, and under the "deictic anchoring", it is in the present. This ambiguity was already noted in antique grammar: *verba dicendi vel sentiendi* "verbs of speaking and thinking" can create their own temporal anchor. This problem is perhaps of limited interest, since it keeps the main idea of temporal anchoring in relation to the speaker's (or thinker's) "now". A tense which related the situation to some anaphorically given time span, as in this example, is sometimes called "relative tense", in contrast to "absolute tenses", in which the temporal anchor is deictic (Comrie 1985).

Second, the speech event itself takes time. How long is the speech time – is it the time it takes to utter the sentence? Or the clause, if the sentence is complex? Many deictic words require a speech time shorter than that time. Consider, for example, the utterance *From now, it is precisely three seconds to now*. (see Kratzer 1978 for this and similar examples). Does it have one or two moments of speech? The conclusion seems obvious: We must distinguish a speech time that is decisive for the tense marking of some sentence from several speech times that are decisive for deictic adverbials within the same sentence.

A third problem shows up, as we consider longer stretches of discourse. Does each sentence in a text, for example a in personal narrative, have its own speech time? Many texts follow the 'principle of natural order', which says 'Unless marked otherwise, order of mention corresponds to order of events'. It is this principle which explains why sequences such as *He fell asleep and turned the light off* are slightly odd. Now, such a principle only makes sense under the assumption that there is a sequence of speech times WITHIN a text, none of which needs to be the actual time at which these sentences are uttered. Hence, it appears that we have to replace the simple notion of 'speech time' by something like a 'structure of speech times', i.e., a set of temporally related time spans which are characterised by particular properties, for example the property that someone says something, or writes something, or hears something.

A fourth problem with the notion of speech time has to do with this characteristic property. It is not always the property that someone says (or writes or hears or reads) something at that time. Consider a sentence such as the following one, taken from the German penal law:

(3)   § 9. [Ort der Tat] (1) Eine Tat ist an jenem Ort begangen, an dem der
            Täter gehandelt hat oder im Falle des Unterlassens hätte handeln

> müssen oder an dem der zum Tatbestand gehörende Erfolg eingetreten ist oder nach der Vorstellung des Täters eintreten sollte.
>
> [approximately: Place of deed. (1) A deed has been committed at that place at which the doer has acted or, in case of omission, should have acted, or at which the result belonging to the definition of the deed has occurred or should have occurred according to the plan of the doer]

This sentence contains many tense forms, *ist begangen, gehandelt hat, hätte handeln müssen, eingetreten ist, eintreten sollte* ('has been committed, has acted, should have acted, has occurred, should have occurred'). What is the temporal anchor of this sentence? It is surely not the time at which the law, or this particular sentence, was (or is?) issued, or at which the reader reads it. Still, such usages of tense forms are by no means exotic (see, for example, the examination of tense in varying text types in Hennig 2000). Hence, the notion of speech time – if this notion is meant to be the time at which the sentence is uttered – is only a special case of how events can be hooked up in time.

These considerations suggest that the notion of "time of utterance" should be replaced by the more general notion of a CLAUSE-EXTERNAL TEMPORAL STRUCTURE, to which situations described by a sentence can be linked. It consists of a set of clause-external times that can be characterised in different ways. Such a clause-external time can be the time at which the entire utterance or a part of it (as in the case of *now*) is uttered or heard; it can also be some other contextually given time. In subordinate clauses, for example, it can be the time of a matrix verb, especially if this verb is a *verbum sentiendi vel dicendi*. Then, ambiguities may arise because there are several possible clause-external times, to which the event can be linked. The interpretation of the 'clause-external time(s)' may vary from sentence to sentence, from text type to text type, and from language to language. Familiar notions such as "moment of speech" or "time of the utterance" are only a special case of such an external temporal structure: it is that type of external temporal structure which we normally use when we talk about single events in reality.

2.4. Non-canonical usages

Tense forms are often used in a way which clearly goes against their "normal" meaning, for example when a present tense form is used to describe events that are clearly in the past. Some of these usages have given rise to extensive discussions, in particular those which are found in narrative fiction; others are less known. Here is a list of such "non-canonical" usages.

A. Narrative present

(4) Two days ago, I walked down the Hauptstrasse in Heidelberg. All of a sudden, a young man looks at me, grins and says: "Hey, don't you remember me?!"

In the narrative present, the whole action is in the past – here indicated by the temporal adverbial *two days ago* -, but that at least some of situations are presented "as if they were present". There are two common interpretations of this use: the situations are "felt to be present" at the time of utterance, or the speaker imagines himself to be present in the situations. Under the first interpretation, the situations are somehow "shifted in time", and under the second interpretation, the deictic anchoring is shifted.

B. Time travel

(5) We are in the year of 2040. The whole world is under the control of three gigantic oil companies. Everybody who counts lives in peace and great luxury.

Here, the first sentence locates the entire action in the future. In other cases, the "time travel" may also go into the real past or a hypothetic past. But the finite verb is marked for present tense. There is no accepted term for this tense use, although it is often found in literary as well as in non-literary texts. Obviously, it is quite different from "narrative present" in the sense of A. above: there is no particularly vivid presentation of the events or feelings of those who participate.

C. "Imagine prefixing"

(6) Imagine you are in a desert. It is very hot, no water around, no oasis in sight. All of a sudden, you hear someone say: "How about a glass of Montrachet?"

This is another type of fictitious discourse. Whereas the "time travel" use pretends to be real and explicitly specifies the time by a calendaric information, this is not the case here: the hypothetical nature of what is said is explicitly marked in the first utterance, a marking which then extends over all subsequent utterances. All of them are in the present, although the first sentence does not necessarily mean *Imagine you are in the desert right now*. In fact, it is very difficult to relate the situations, as described in the text, to the time of utterance.

D. Praesens tabulare

(7) In 1819, Goethe publishes "Die Wahlverwandschaften". They are completely ignored by the critics. He is deeply disappointed and almost decides to give up literary work.

In this use, a series of historical facts is simply registered, and it is clearly stated when these facts happened – and they are in the past. There is no mental "moving", no fiction nor any vivid narration whatsoever, quite to the opposite: It is a sober presentation of situations in the past.

E. Epic preterite

Since this case seems more common in German than in English, I give a German example here:

(8) Er wanderte durch die stillen Strassen. Morgen war Weihnachten. Niemand würde ihn erwarten, niemand würde ihn vermissen.
['He was wandering through the quiet streets. Tomorrow was Christmas. No one would wait for him, no one would miss him.']

The most typical tense of literary narration is the preterite. But the situations are not "really" in the past. They are in some hypothetical time. Moreover, it is not very clear what the "time of utterance" of, say, a novel is. Some authors have concluded that the tense form in this case does not express a tense relation but a discourse type (Hamburger 1956).

F. Re-telling

(9) In the next scene, Charlie looks around everywhere. Then, he discovers an old shed and begins to walk there.

This is a common although not the only possible tense use in re-telling a movie (or a story). The action is clearly not in the present. In fact, it is not very clear how the action can be related to TU at all. It is a movie, and what happens in the movie is not part of the chain of events or states which eventually lead to the present and the talking about them. But the action is presented as if it is just seen at the time of speaking.

G. Past in pictures

(10) This is Eva when she was four years old. And here, she just got this little bike. She looks very cheerful, doesn't she. This was in 2002.

The picture shows an action which is long ago (or was long ago). But it is shown right now, and it is apparently this time which the speaker chooses to talk about, rather than the time at which the action really happened. He need not choose this "time of picture looking" as topic time, as the subordinate clause in the first utterance, the second and the last utterance show. But he can, and he may switch even in such a short stretch of discourse.

H. Backchecking

This use is normally not observed in longer stretches of discourse but in short questions in some situations, for example:

(11) a. (Waiter): Who got the Chardonnay?
b. (To visitor): Sorry, what was your name?

Here, the use of the past tense form surely does not refer to the past of the relevant situation.

This list of "non-canonical usages" is not exhaustive. But it suffices to show that the idea of a stable relation between, for example, pastness marking and pastness is a bit of an illusion.

2.5. Summary

The old idea that tense is the main device which languages use to encode time is the result of several strong biases. The classical definition, under which tense indicates whether the situation described by a sentence is in the past, the present, or in the future – in other words, whether it precedes, is simultaneous to, or follows the speech event – works only in special cases, and many languages have no tense marking on the verb at all. This does not mean that tense is unimportant; but its role for the expression of time may be a bit overrated in the research tradition.

3. Aspect[3]

3.1. The general idea[4]

Like tense, aspect is an inflectional category of the verb; unlike tense, it does not assumed to express a temporal relation but a particular "viewpoint" on the situation which is described by the sentence. The speaker may present the time-course of the situation in different ways. The most common distinction is between an "imperfective" aspect and a "perfective" aspect. In the first case, the situation is presented as on-going, as in English *Mary was opening the window*; it need not have come to an end. In the latter case, it is presented as completed, as in *Mary closed the door*. There are other possible viewpoints (see, e.g., Comrie 1976), but perfective and imperfective are by far the most important. The term comes from the French translation *aspect* of the Russian word *vid'* "view" (Reiff 1828/9). Russian, as most Slavic languages, has a very salient marking of two such "views" in the verbal system, and in the early 19th century, linguists realised that the conventional description in terms of tense alone does not do justice to the facts. Although Russian may not be the most typical case of a language with such a grammaticalised view (see Dahl 1985), the notion and also the basic distinction between perfective and imperfective made their way into the analysis of many other languages.

During this process, many attempts were made to render the idea of different temporal views on a situation more precise. Numerous characterisations are found in the literature; they can be grouped into three types:

(12) A. The situation is presented "from outside" versus the situation is presented "from inside".

B. The situation is presented as "completed" versus the situation is presented as "non-completed" or "on-going".

C. The situation as presented "with its boundaries" versus the situation is presented "without its boundaries".

---

[3] The term "aspect" is also used as a label for situation types, such as states, processes, events, sometimes with the qualification "lexical aspect" (in contrast to "grammatical aspect" or "situation aspect" in contrast to "viewpoint aspect" (Smith 1997). In order to avoid terminological confusions, I will use it here only in the classical sense.

[4] A very clear introduction is still Comrie (1975). Dahl (2000) and Ebert and Zuniga (2001) are very useful reviews of aspect systems in European and Non-european languages, respectively.

It is easy to see that these characterisations are not necessarily exclusive; they are somehow variations on the same underlying theme, and they are not entirely clear. Before coming to various problems connected with them, it will be useful to have a brief look at two well-investigated cases, English and Russian. They exemplify different ways to encode aspect and to combine it with tense.

3.2. The case of English

The English system is remarkably regular in both respects. Suppose the situation is described by [Eva sleep]. Then, this situation can be related to the time of utterance TU[5] ("tense") and presented under different viewpoints ("aspect") in the following ways:

	perfective	*imperfective*
before TU	Eva slept	Eva was sleeping
at TU	Eva sleeps	Eva was sleeping
after TU	Eva will sleep	Eva will be sleeping
before TU	Eva has slept	Eva has been sleeping

Transparent as this system is – it still raises some smaller and at least two major problems. A few verbs (*to need*) do not tolerate the *be -ing*-form, or they only tolerate it in a special meaning (*to love*). These verbs are typically stative; but in the course of development, most stative verbs came to take the progressive (Denison 1993: 371–410). While tense and aspect are in principle orthogonal, there are some restrictions on the use of forms such as *Eva sleeps*. These restrictions may have to do with the precise meaning of this tense-aspect combination, and thus with the first major problem: what exactly is the meaning of these forms? This concerns the tense side as well as the aspect side. There are two "tenses" which place the situation before the time of utterance – the simple past *slept* and the present perfect *has slept*. The difference is clearly felt, and it is manifest in facts such as that in the present perfect, the time of the location cannot be specified by an adverbial (**Eva has slept yesterday at four*). But it is not easy to pin down this difference in terms of tense and aspect. In both cases, the situation is in the

---

[5] Since this is only for the sake of illustration, the many complications with tense discussed in the preceding section are ignored here.

past, so, they both should be past tenses. They both occur in both aspects, so, the difference cannot be in terms of "imperfective" and "perfective". It could, of course, be that there is a third viewpoint, call it "perfect". This idea is supported by the fact that on closer inspection, the perfect forms constitue a tense system on their own, based on the auxiliary marking:

	perfective	*perfective*
before TU	Eva had slept	Eva had been sleeping
at TU	Eva has slept	Eva has been sleeping
after TU	Eva will have slept	Eva will have been be sleeping

But if the perfect is also an aspect, it cannot be on a par with the perfective and the imperfective aspect, because these are found **within** the perfect; these latter would, so to speak, be viewpoints within a viewpoint. There is a very rich literature on the English perfect (see Fenn 1987; McCoard 1978; Klein 1994; Comrie 1985; Portner 2003; and for comparisons to other "perfects" Musan 2002 and the contributions in Alexiadou et al. 2003), but one cannot say that there is a generally accepted analysis.

This problem leads us immediately to the second major problem of the English tense-aspect system: what exactly is the meaning of the two aspects? The English variant of the imperfective is often called "progressive" or "continuative", and this makes perfect sense in cases such as *Eva was frying two eggs*: we are somehow placed in the midst of the action, it is presented as on-going, as proceeding, with reference to "its inner temporal constituency" (Comrie 1976). In the perfective form *Eva fried two eggs*, we have the impression that the action is somehow completed, that it has reached the end. But this is much less clear in contrasting pairs such as *It was standing on the marked place* vs. *It stood on the market place* or *Soames was hoping for a rapid solution* vs *Soames hoped for a rapid solution*. In neither case is there any progress, any inner temporal constituency. This is just a reflex of a much more general problem – how can the definitions given above under A – C be made precise, and in which way do they vary from language to language? We will come to this in a moment.[6]

---

[6] As an aside, it is not encouraging for the working linguist when even in the best-studied language of the world, such as salient phenomenon like the *-ing*-form is not really understood.

3.3. The case of Russian

Russian shows a much more complicated interlace of tense and aspect. The basic facts can be summed up in seven points (there are a number of exceptions and complications, for example loan words or some special motion verbs that are ignored here; an excellent and comprehensive survey is Forsyth 1970, more generally on Slavic aspect Dickey 2001):

1. Each verb belongs to one of two aspects, the imperfective or the perfective.
2. As a rule, morphologically simple verbs are imperfective, e.g. *pisat'* "to write"; there are about 30 simple verbs which are perfective, e.g. *dat'* "to give"; a few verbs are ambiguous.
3. Attaching a prefix to an imperfective verb makes it perfective, e.g., *na-pisat'*; there are about 20 such prefixes.
4. At the same time, this prefix typically modifies the lexical meaning of the verb to some extent, e.g. *na-pisat'* "to write (up)", *pere-pisat'* "to copy". Sometimes, this change is substantial, sometimes, it is minor (or non-existent, as claimed by some authors).
5. Many but by far not all verbs that are made perfective in this way can be made imperfective again by adding a suffix to the stem (so-called secondary imperfectivisation). So, we have the chain *pisat'* – **perepisat'** – *perepisovat'*. Such a suffix does not change the lexical meaning – it indicates is a bare aspectual contrast.
6. The present tense form of an imperfective verb has present tense meaning; the present tense form of a perfective verb has future tense meaning. The future meaning of an imperfective verb is expressed by an analytic construction: *ja budu pisat' pismo* "I will write (a/the) letter".
7. The past tense meaning of perfective as well as imperfective forms is expressed by attaching a suffix -l to the stem; this suffix is inflected for gender and number (historically, it is a sort of past participle, but there is no remnant of an auxiliary): *ona pisala pismo* "she was writing a letter", *on perepisal pismo* "he copied a/the letter".

It was this system which originally gave rise to the notion of grammatical aspect; but apparently, Russian aspect operates in a very different way than aspect in English. First, there is a somewhat idiosyncratic combination with tense meanings. Second, the way in which aspects are marked is very different. Third, prefixation usually changes the viewpoint as well as the lexical meaning of the verb; this is not true for secondary imperfectives, which have the same lexical meaning as the perfectives from which they are derived. And fourth – it is not clear what exactly the bare aspectual meaning is. Although characterisations such as given in (12) are used for Russian as well as for English, there are substantial differences. Thus, Russian "imperfectives" often do not indicate on-goingness, and in English, they would have to be translated by the simple form rather than by the progressive. In nutshell: "imperfective" in language A does not mean "imperfective" in language B.

I have chosen these two examples of tense-aspect systems, because they belong to the most-studied of all languages of the world, dead or alive. They demonstrate that "aspect" is neither in form or meaning a homogeneous phenomenon. In what follows, we shall address some of the core problems in more detail. They all have to do, in one way or the other, with the way in which aspect and aspects are is defined.

3.4. Three problems

3.4.1. What is a "viewpoint"?

Characterisations such as in (12) are found everywhere in the literature, and they are very suggestive indeed. But they are entirely metaphorical. What exactly is meant by "to see/view/present a situation in different ways"? The situation itelf is supposed to remain the same when different aspects are chosen: it is only the perspective on it that changes. But it is hard to make precise what this metaphor means. Most languages allow their speakers to express a particular content relative to their position, and thus from their "perspective". A good and clear example is deixis; depending on the position and gaze direction of the speaker, one and the same constellation can be described as *here* or *there*, as being *to the left* or *to the right*. But this is surely not what is indicated by the choice of a particular aspect. In contrast to tense, aspect is not deictic. Situations such as the ones described by [Mary bake a cake] or [Peter sit on an old chair] are not like chicken shacks, which you can "see" from the inside and from the outside.

3.4.2. On-going when, completed when?

Under the traditional definitions, aspect is not time-relational: in contrast to tense marking, aspect marking is not supposed to relate the situation to a particular time. But in fact, it is time-relational: on-goingness as well as completion are always relative to a particular time. Consider, for example, the situation which is described by the English sentence *Eva boiled an egg*. If this sentence is true, then the situation referred to is completed, for example, at 6 o'clock, and it is not completed at 5:45 o'clock. At that time, it may be on-going, or it may not have begun at all. Therefore, a definition like "The perfective aspect presents a situation as completed" makes sense only if it means: "the situation is presented as completed **at some time T**". A speaker who presents some situation as completed does not want to suggest that it was or is completed at any time: It is completed at some time T, as well as at any time thereafter, and is not completed at certain times before T. The notion of completion crucially depends on "the time about which something is said" – for example 5:45 o'clock or 6 o'clock in this example. This time need not be made explicit; in particular, its relation to to the time of utterance need not be expressed. What is this – possibly implicit – time, in relation to what is said to be completed or on-going? Without an appropriate definition of this notion, the entire characterisation as 'presented as completed – not completed' is hanging in the air.

3.4.3. Which boundaries?

The third problem has to do with the notion of boundary, found in many characterisations of aspect. With very few exceptions, all situations have a beginning and an end – an "initial (or left) boundary" and a "final (or right) boundary". This does not mean that the **description** of a situation by the sentence makes these boundaries explicit. The speaker may choose to mention them or to leave them aside, just as he may mention or not mention other features of the situation. On-goingness and completion seem naturally related to these boundaries, in particular to the "right boundary". Therefore, it is suggestive to assume that aspect is a device to make these boundaries explicit. Take, for example, the following definitions by Carlota Smith, which are particularly clear in this regard: "The perfective viewpoint ... presents events with both initial and final endpoints." (Smith 1997: 301) and "The temporal schema of the imperfective viewpoint focusses on part of a situation, excluding its initial and final endpoints." (Smith 1997: 302).

Inutitively, such a view is very appealing; but there are at least two reasons which render it unsatisfactory. Suppose we describe a situation by [Eva sleep from four o'clock to six o'clock]. Then, the boundaries are clearly indicated by the two temporal adverbials. Nevertheless, we are free to present the situation so described in the imperfective and in the perfective aspect: *Eva was sleeping fom six o'clock to eight o'clock* vs *Eva slept from four o'clock to six o'clock*. So, this choice of aspect seems – at least in these cases – independent of whether initial and final boundary are "visible" or not. The second problem is that the characterisation of different Aktionsarten – states, processes, events – is often based on the presence or non-presence of a boundary. In Vendler's widely used classification, four such "time schemata" (as he calls them) are distinguished: states, activities, accomplishments and achievements. Verbs (or verb phrases) which describe a state or an activity, such as *to stand* or *to run*, do not involve such a boundary, whereas verbs (or verb phrases) which describe an accomplishemt (*to paint a picture*) or an achievement (*to find a solution*) are inherently bounded – bounded due to their lexical meaning. Now, if the semantics of grammatical aspect is defined in terms of boundaries, as well, then the difference between inherent lexical properties of the verb, on the one hand, and aspect, on the other, is entirely confounded. If the perfective aspect somehow involves a boundary, then this boundary must be of a different type that the boundary inherent to the lexical content of the verb. In Russian, verb pairs such as perfective *dat'* and imperfective *davat'* 'to give' or perfective *perepisat'* and imperfective *perepisyvat'* "to copy" (cf. section 3.3 above) are said to have exactly the same lexical meaning; in Vendler's terms, both would be accomplishments, hence involve some inner boundary. But they differ in aspect. Hence, the perfective aspect should add some other, additional boundary. What is this boundary, and how does it relate to the boundaries of the situation type?

3.5. Summary

Aspects are different ways " to view" or "to present" one and the same situation. Many languages have grammaticalised such a distinction, typically in form of a "perfective" and an "imperfective" aspect; some languages are also claimed to have other types of viewpoints. This notion captures an important intuition about how time is encoded. But it is not very clear. The definitions which are typically used, as the ones in (12), are metaphorical, and they miss important facts, in particular the fact that completion and non-completion are time-relational. This asks for more precise definitions, and in fact, several attempts in this direction have been made in recent years – often in combination with treatments of tense (for example Smith 1997; Klein 1994; Giorgi and Pianesi 1997; Declerck et al. 2006). But by far most descriptive work on form and function of aspect in natural languages is based on these intuitive characterisations.

4. Aktionsarten

4.1. Lexical features

Tense and aspect are grammatical categories of the verb. But temporality is also reflected in the lexicon – in the content of verbs and other lexical items which serve to describe situations. Situations themselves vary considerably in their inherent temporal structure. There are rapid events, such as the explosion of a bomb, there are complex actions such as the baking of a cake, there are gradual and slow processes as the melting of an iceberg, there are states such as Eva's being cheerful, and there are even "atemporal" situations such as the fact that 17 is a prime number. These differences can be encoded by the words which describe them, and thus, we have different types of verbs and other expressions. The oldest and perhaps still most-used traditional term for these types is "Aktionsarten" (manners of action), and we will adopt it here; others are "event types", "lexical aspect", "situation aspect".

Aktionsart distinctions are found in all languages. Traditionally, their primary source is the lexical meaning of the verb; there are "event verbs" such as *to explode*, there are "action verbs" such as *to speak*, and there are "state verbs" such as *to hope*. But is easy to see that the temporal characteristics of the situation which the sentence describes vary with many other words, for example adverbials or the direct object. Thus, there is a clear difference between *to smoke*, *to smoke for half an hour*, and *to smoke three cigars*. In fact, all words in a sentence can contribute by their lexical meaning to describe the temporal make-up of a situation. In what follows, however, I will concentrate on verbs and verb phrases, since these are the expressions which are normally investigated in this context.

What are the inherent temporal features of these expressions? There are innumerable attempts to answer this question, beginning with Aristotle's distinction between "verbs of kinesis" and "verbs of energeia"[7] to some at-

---

[7] His definition is as follows (Metaphysics, Theta, 6, 1048b): "Thus, you are watching and thereby have watched already, you are thinking and thereby have thought already; by contrast, you are learning (something) and have not learned (it) already, and you are becoming healthy and are not yet healthy. At the same time, we are living well and have lived well, we are happy and have been happy. Otherwise, the process should have ended at some time, like the process of becoming thin. But it has not come to an end at the present moment: we are living, and have lived." (for a discussion of the tradition, see Taylor 1965).

tempts to describe these features in terms of formal semantics, such as Dowty (1979) or Krifka (1989). They range from two or three categories to very complex systems; Noreen (1923), for example, distinguishes 17 Aktionsarten. Nowadays, most researchers use Vendler's (1957) four classes state, activity, accomplishment, achievement, or they develop their own classification, often tuned to the particular language they want to analyse. Different as these systems are, they are all more or less based on the following five temporal features:

(13) A. Qualitative change: does the content which is expressed involve a change of state or not (non-stative vs. stative VPs)?
  B. Boundedness: does the content which is expressed have a beginning and an end, or, as is often said, an initial and a final boundary ("unbounded" vs. "bounded", often contrasted as "processes" vs. "events")?
  C. Duration: in the case of "bounded contents", are they short or long in duration ("punctual" vs. "non-punctual" contents)?
  D. Inner quantification: do they involve repeated sub-events or sub-states ("iterative", "frequentative", "semelfactive")?
  E. Phase: do they focus on a sub-phase of the total content, for example the beginning, the middle, the end ("inchoative", "terminative", "resultative", etc.)?

In what follows, I will not discuss these five temporal features individually but concentrate on two general problems which these features and the classifications based on them raise.

4.2. Temporal properties of the situation and of its description

The lexical content of a sentence, for example [Eva sleep in the guestbed] or [Eva work in the garden], is a description of a situation (or set of situations). This description is selective: the situation may have many features, temporal or other, that are not part of the description. In reality, almost everything has a beginning and comes to an end, be it a state, a process, an event; there are only a few exceptions, such as (probably) *Seven is a prime number*. Hence, there is hardly any situation which does not have an initial boundary and a final boundary. But this does not mean that each verb or verb phrase must include such a boundary as a part of its lexical meaning. Therefore, a distinction is often made between "unbounded expressions", such as *to stand on the table* and "bounded" expressions, such as *to put on the table*; the latter somehow involve a terminal point – point at which the "putting" as such is over and a cup is on the table. Similarly, it is usually assumed that, while there are no really punctual situations in reality, languages may conceive of situations as having no duration, and thus being punctual. This is assumed to be the case for verbs such as *to find*. It is this sense in which we speak about "punctual verbs".

The distinction seems clear-cut: We do not talk about what is the case in reality but about the way in which languages grasp and encode reality in lexical contents. But this is not so straightforward. Consider, for example, the situations described by (14):

(14) a. The cup stood on the table.
  b. Eva put the cup on the table.

In the normal course of events, both situations have a beginning, an end, and thus a duration; we do not assume that the cup's standing on the table, as described in (14a), lasts forever, just as little as her putting it on the table lasts forever. What, then, does it mean that a lexical item has a temporal feature like "having duration" or "having a right boundary", if it does not mean that the situations typically described by VPs such as *to stand on the table* or *to put on the table* do have a duration or a right boundary? Any situation described by a "state verb" such as *to lie*, or by an activity verb such as *to snore* takes some time, hence may be long or short; the lexical content of these verbs themselves has no time, there is no watch to measure its duration, and if we imply that it has duration, then only by virtue of the fact that the situation it refers to can be measured by a watch. There is no time at which the lexical content can ever be over, because it has no time at all. What can be over at some point in time, are the situations to which we may refer by means of these words. If we say that a certain lexical item involves boundaries, then it is only by virtue of the fact that this lexical item can be applied to situations which indeed have a duration, or an end. Hence, it seems mysterious that a set of situations should have duration, or an end in reality, but the corresponding lexical contents do not contain such a temporal feature. But if this is true, the distinction between "event VPs" with and "state VPs" without boundaries, or between punctual and non-punctional verbs (or VPs) collapses, because in reality, there are no situations without duration. This, however, means that virtually all known systems of verb classification are on shaky grounds – not for practical but for principled reasons.

4.3. Semantic intuition vs morphosyntactic combinations

Classifications of verbs into various Aktionsarten are usually based on two methods:

A. Semantic intuitions; thus, someone who knows English also "knows" that *to lie on the table* is something static and does not involve a change, whereas *to put on the table* is something dynamic and involves a change: first, the object of putting is not on the table, and then, it is.

B. It can also be based on the way in which verbs are affected by morphological or syntactic operations, for example how they interact with temporal adverbials.

Now, semantic intuitions are often fuzzy, as becomes immediately clear when we run through a series of examples. I shall not illustrate this here but encourage the reader to go through any bit of text and try to determine the temporal properties of each verb (for example the verbs in this sentence). The second method seems more reliable, and has therefore found wide application in the research tradition. The following three types of innerlinguistic tests are often used:

(a) Modification by adverbials: it is tested whether the VP in question can be combined with a specific adverbial, for example *he VP in two hours, he VP for two hours*. A variant of this way is to check whether a specific wh-question is possible, like *How long did it take to VP?*

(b) Aspectual modification: it is tested whether the item in question is accessible to aspect modification. The best-known case is the "*-ing* test" for statives in English. It is argued that stative verbs like *to contain, to belong, to know* do not tolerate the progressive, hence it is easy to test whether a VP is stative or not.

(c) Presuppositions and implications: It is checked whether a sentence containing the crucial form has a certain presupposition or implication. This test was already used by Aristotle (see fn. 7). A more recent version was proposed by Garey (1957) in order to distinguish between "telic" and "atelic" verbs: "If someone was V-*ing*, then he V-*ed*" (for example, *If someone was washing the car, then he has washed the car* vs *if someone was living in London, then he has lived in London*); only atelic verbs pass this test. Another example, first suggested by Bennett and Partee (1978), is the test for the sub-interval property: If something is true for some interval t, is it then also true for any sub-interval of t?

All of these methods have been extensively applied, and they are probably the best available diagnostic for verb classes; thus, Vendler's famous classification is entirely based on a combination of (a) and (b): states and activities go with *for two hours*, whereas accomplishments and achievements do not; activities and accomplishments take the *ing*-form, whereas states and achievements do not.

This way to proceed is more satisfactory than a mere appeal to semantic intuitions. But there are also some problems. Consider, for example, aspectual modification. There are, first, many languages in which it cannot be applied, since they have no grammatical aspect; this raises serious questions as to the transferability of, for example, Vendler's four types to other languages (a fact which so far has hardly prevented anyone from transfering them). But even if it can be applied, as in English, its validity hinges on the assumption that stative verbs (or VPs) do not tolerate the imperfective. Now, locative predicates such as *standing on the table* admit the *ing*-form. Therefore, the claim is either circular (if "stative" is **defined** as "not tolerating the imperfective"), or else, it is false. The problems are less obvious with the two other methods; still, there are a number of questions to be asked.

The difficulty with (a) is simply that it is very often not clear what is really modified. I will illustrate this with Vendler's "for x time" test, since it is so often used in the literature. Some VPs do not tolerate an adverbial like *for two hours, for ten seconds*, etc, whereas they tolerate modification by *within one hour, within four seconds*. Vendler uses this test to separate "achievements" and "accomplishments", one the one hand, from "states" and "activities", on the other: one can say *He reached the station in one hour*, but not *He reached the station for one hour*. Hence, *to reach the station* is an achievement or accomplishment; for VPs such as *to stand on the table* or *to be in Heidelberg*, it is exactly the other way around. This is a useful test, without any doubt, and in contrast to the aspect change test, it is better transferable to other languages. Now, the exclusion of *for*-adverbials is not absolute; thus, *She opened the window for two hours* is possible – though only, if the adverbial indicates the duration of the resulting state, rather than the duration of the opening-activity. This raises two connected questions: first, what exactly is specified by the adverbial, and second, what is the reason that the duration cannot be specified in some cases? Under the usual analysis of *She opened the window*, there is a right boundary (the point in time at which the window is first open), and it is also clear that it took some time to open the window, say three seconds, one minute, or even half an hour (some windows are hard to open). Hence, the situation lasts for a while, although perhaps for a short while only; it is not punctual. Why should it then be impossible to use a durational adverb with it, just counting the time of when his opening began until the first moment at which the window was open? Therefore, the constraint exploited by the adverb test seems a complete mystery.

I think the reason is that the lexical content of *to open a window* involves neither a "right boundary" nor punctuality. It simply includes two qualitatively distinct states – a "source state", in which the window is not open, and a "target state", at which it is open; during the first state, there is also some kind of activity on the part of the agent (for example, turning a handle etc). The lexical content does not say anything about the length of these states; all it implies is that there is a transition between the source state and the target state, which may be abrupt or smooth; otherwise, it would not be possible to talk about two qualitatively distinct states. A durational adverbial such as *for two hours* cannot modify both states at the same time, since they are mutually exclusive; at best, it can apply to the "resulting state", and this indeed yields a possible reading of *He opened the window for five minutes*.

This does not speak against an application of such tests – but it speaks against a blind application. In each case, we must consider, what is exactly modified by a particular adverbial, or by some other kind of morphosyntactical operation, and why it can be applied or not.

5. Temporal adverbials

Across all languages, temporal adverbials constitute the by far most elaborate device to encode time. They vary considerably in form and function. In what follows, we illustrate this for English, which stands here for all Indoeuropean languages.

5.1. Forms

English has essentially the following three types of temporal adverbials:

A. Simple temporal adverbials: *now, then, soon, often, seldom, just*.
B. Morphologically compound adverbials: *today, afterwards, sometimes, slowly*.
C. Syntactically compound adverbials. This is the richest class, with three main constructions:

a) Bare noun phrases: *last fall, all day long*
b) Prepositional (or postpositional) phrases: *three hours ago, three hours before, after the autopsy, in the past, for seven years, at any moment.*
c) Subordinate clauses: *before he arrived, while I was in China, whenever she called me.*

They often combine to more complex constructions, for example: *soon after the autopsy, at four o'clock on every second Friday in 2004* or *before and sometimes after he was in Riva*.

5.2. Functions

Since they can draw on virtually all lexical and grammatical means of a language, temporal adverbials allow highly a differentiated characterisation of all sorts of temporal features. Essentially, four functional types are found.

A. Temporal adverbials of position

In English, these include, for example *now, two days ago, after the riots*. Temporal adverbials of position express a relation such as BEFORE, AFTER, SIMULTANEOUS between two time spans – a time which somehow positioned (the "theme"), and a time which is used as an anchor, in relation to which the theme is positioned (the "relatum"). In *Max will arrive soon*, the adverbial *soon* expresses a temporal relation AFTER between the time of a situation, described by [Max arrive], and the time of utterance. The adverbial *in five minutes* in *Max will arrive in five minutes* expressed the same type of relation: theme AFTER relatum, but it makes the distance between the two times more precise. Note that in both cases, the temporal relation is also indicated by the tense form *will arrive*. This information is here completely redundant.

In this example, the relatum is deictic – it is the time at which the sentence is uttered. There are two other important types of relata – anaphoric and calendaric. Anaphoric relata come from the preceding context, as in *Then/a few minutes later/after the call/after the had closed the shop, he left*. Calendaric relata use some historical event as an anchoring point, such as the birth of Christ: *In 2002, he died*. Note that the relatum itself is not made explicit in this case; what this sentence means, is: "in the year which is 2002 years after the calendaric relatum, he died".

B. Temporal adverbials of duration

The duration can be indicated in a vague way, as in as *He worked for quite a while, The exam lasted forever*. It can also be made very precise, as in *He worked for seven hours and four minutes*; in this case, the theme (the time of the situation whose duration is specified) is related to multiples or fractions of the duration of some other situation, for example the rotation of the earth around its axis or around the sun.

Note that adverbials such as *soon* or *three days ago* also have a durational component. Their primary function is to indicate a temporal relation between two time spans, such as AFTER or BEFORE. In addition, the indicate the duration of the time **between** these two time spans, either in a vague (*soon*) or in a more precise way (*three days*).

C. Temporal adverbials of frequency

They quantify over time spans. As in the case of durational adverbials, this can be done in a relatively vague way, such as *often, sometimes, on several occasions*, or more precisely, as *twice a week, every Friday*. A particularly intesting case is the adverbial *always*, which seems to quantify over all times; so, it seems to mean "at all times". In actual fact, however, it is normally used to express the idea: "at any relevant time" and thus carries a flavour of subjectivity. Thus, a sentence such as *He always forgets to take his teeth out* does not mean "For all times, it is true that he forgets to take his teeth out" but "At any time when it would be relevant for him to take his teeth out, he forgets it".

D. Temporal adverbials of contrast

All languages we know of also have a type of adverbial that does not specify features such as relative position, duration or frequency but still have a clearly temporal flavour. Typical examples in English are words like *still, already*, and *again*. Their precise function is not easy to pin down. Compare the following four sentences:

(15) a. Eva was in Riva.
 b. Eva was already in Riva.
 c. Eva was still in Riva.
 d. Eva was again in Riva.

The first sentence states that some time T in the past (the "topic time") overlaps with the time of a situation described by [Eva be in Riva]. When *already* is added, then this indicates that an earlier but adjacent time also has the property [Eva be in Riva]; so, (15b) means: at some time T, she was in Riva, and at some time immediately before T, she was in Riva, as well. The addition of *still* has a similar effect, except that it "adds" a later (and adjacent) time to T. The meaning contribution of *again* is something like "and this not for the first time". Thus, (17d) means: "At some time T, John was in Riva, and at some time before T – which must not be adjacent –, he was in Riva, as well."

In these examples, it is relatively easy to describe the function of these adverbials. There are, however, many complications. Most of these are due to varying scope. This is particularly clear for *again*. It was often noted that English again can have two readings, as exemplified by (16):

(16) Eva opened the window again.

This can mean that Eva opened the window, and this not for the first time ("repetitive reading"), or that she opened the window and thereby restored an earlier state [window be open] ("restitutive reading"). As simple way to account for this ambiguity is to assume that verbs such as *to open* include two distinct times: a "source state" at which the agent does something and the window is not open, and a "target state" at which the window is open. An adverbial such as *again* can have scope over both times, and then, the entire action is said to be repeated, or it can have scope only over the "target time", and then, it is said that the target state does not obtain for the first time. In other languages, such as German, there are many other scope possibilities which allow for various types of repetition or continuation (for a discussion, see Fabricius-Hansen 2001; Klein 2001; von Stechow 1996).

This brings us to the next point – the way in which adverbials interact with the remainder of the sentence.

5.3. Interaction

Under a very simple view, temporal adverbials characterise the position, duration and frequency of the entire situation described by the sentence; Thus, the initial adverbial in *Yesterday, Eva left.* indicates that the time of situation described by [Eva leave] is at some time within the day before the day which includes the utterance time. As the example of *again* has shown, the case is much more complicated. Adverbials can be inserted in various ways into the sentence, and accordingy, their effect on the entire meaning can vary considerably. It also varies considerably from language to language.

In what follows, we shall have a look at a few English examples which illustrate this point.

The following sentences have in common that the situation itself, as described by [Eva leave], is always in the past; this is indicated by tense marking, but also by adverbial (I assume that *at five* refers to some time in the past):

(17) a. At five, Eva left.
   b. Eva left at five.
   c. At five, Eva had left.
   d. Eva had left at five.
   e. *At five, Eva has left.
   f. *Eva has left at five.

It is hard to tell the difference between (17a) and (17b); in both cases, we assume that the adverbial *at five* specifies the exact time of her leaving. This is different in (17c) and (17d); if the adverbial is in initial position, it specifies a "posttime" of the leaving situation, and as a consequence, the leaving itself must be earlier, for example at four o'clock. In (17d), the same adverbial gives either the "leaving time" or one of its posttimes; the interpretation depends partly on the intonation: if *at five* carries main stress, it is normally understood to give the leaving time; if *at five* is de-stressed, it gives a posttime. In (17e) and (17f), the adverbial is odd. This makes sense for (17e), because there is a clash between the past adverbial *at five* and the present tense *has*; but it is perplexing why (17f) does not have a reading under which *at five* indicates the time of leaving. This restriction has found some attention under the label of *present perfect paradox* (see, for example Dowty 1979, Klein 1992); but there is no generally accepted answer so far.

In (17), the sentence itself is simple; there are more complex cases, for example (18):

(18) a. Miriam appeared to have planned to open the window.

What is "the situation" in this case, and what is "its time"? Obviously, there are many times – the time at which something appears to be the case, the time at which Miriam appears to plan something, the time at which this planning is over, the time of her intended opening, and maybe still others. What happens if an adverbial such as *at five* is inserted in (18)? This is not easy to tell, since intuitions on which positions make sense and on what they imply are shaky:

(18) b. At five, Miriam appeared to have planned to open the window.
   c. Miriam appeared at five to have planned to open the window.
   d. Miriam appeared to have planned at five to open the window.
   e. Miriam appeared to have planned to open the window at five.

In initial position, the adverbial is normally felt to give the "appear-time", possibly also the planning-time, but surely not a posttime of planning, nor the opening time. In (18c), it is also the appear-time that is targetted by *at five*; all other time spans seem more or less excluded. In (18d), it is the planning-time, and in (18e), it is the opening time, and all other times are excluded – more or less.

In all examples so far, the adverbial specifies the position of some interval on the time line (here by the calendaric expression *at five*). We note the same problems with adverbials of duration or of frequency. I will not go through the corresponding sentences but will have a brief look at a somewhat different example:

(19) a. Twice, one of my colleagues bought a house.
   b. One of my colleagues twice bought a house.
   c. One of my colleagues bought a house twice.

In each case, there are two situations of the type described by [one of my collegues buy a house], as indicated by the adverbial *twice*. But the degree to which these situations are identical varies. In (19a), neither the colleague nor the house must be the same, in (19b), the normal reading is that it is the same colleague but not the same house (although this is not entirely excluded), and in (19c), the normal reading is that the colleague and the house are the same – but there are two "buying actions" at different times. This shows that the position of the adverbial does not only affect which time spans – if there are many – are addressed but also how a particular time span is characterised by the descriptive content of the sentence.

6. Temporal particles

If one were to invent a simple and elegant system to express the classical tense and aspect functions, then one should perhaps form a few simple morphemes, say *tu, ti, ta* for "in the past, at present, in the future", respectively, and *le, lo* for "completed, ongoing", respectively. They should have a clearly defined position (say immediately after the verb), and they should not be mandatory, i.e., leave it to the speaker to express whether something is in the future and whether something is completed; finally, tense and aspect functions should be freely combinable, so we have *tule* for "completed in the past", *talo* for "ongoing in the future", and so on.

In no case has the collective intelligence of all speakers of a language ever created such an elegant system. Human languages are not that functional. But some languages have developed structures which use a bit of this potential. The best-known case is Chinese, which has neither tense nor aspect inflection but uses a few particles which can follow or (in one case) precede the verb. In Mandarin, the most important of these are *le, guo, zhe* and *zai*. Their formal as well as their functional properties are an issue of vivid dispute, but the rough picture is as follows:

A. The particles *le, guo* and *zhe* follow the verb, *zai* precedes it.

B. All four particles express a sort of aspect, in particular:

   (a) *le* indicates perfectivity, independent of the position of the situation on the time line

   (b) *zai* and *zhe* indicate on-goingness, independent of the position on the time line

   (c) *guo* indicates that the situation has been experienced at least once in the past.

C. None of these particles is mandatory, although sentences without them often sound somehow "hanging in the air", that is, they are not really embedded in the communicative context.

There are many complications, but this may suffice to illustrate the general idea (for a more detailed discussion, see Smith 1997; Klein, Li Ping and Hendricks 2000; Xiao and McEnery 2004). Few languages proceed in this way. This is somewhat surprising since such a way to encode temporality – when combined with adverbials – would be simple, versatile, and it would give its speakers a lot of freedom.

7. Discourse principles

At the end of section 1, it was pointed out that research on the expression of time suffers from a one-sided bias towards singular events. But even a singular event has typically a very complex temporal make-up, which cannot easily be described in a single sentence. Suppose you are asked to report on a car accident that you happen to have witnessed. What you should talk about, is a complex event that took place at a certain time at a certain place and that has a very rich inner temporal structure: it consists of numerous sub-events which take a certain time and which may be more or less simultaneous or sequential. In other words, you must encode an extremely complex web of temporal features. Tense and aspect may help here, but it is easy to see that they are of very limited use. Once it has been said that the accident was in the past – in fact, it probably need not even be said in such a communicative situation -, tense is almost worthless, and aspect does not really allow you to depict the web of temporal relations. Adverbials (*then, while ..., before ..., for about two seconds*) are more useful, but one of the mightiest devices is to follow certain "strategies" on how to decompose the entire event into smaller sub-events and to report them in a certain order. The report of the car accident includes at least the following tasks on the speaker's part:

A. The entire event must be embedded in time (and space). Typically, this happens by some initial adverbials, often in combination with tense: *Well, all of this happened around four o'clock, and it lasted only a few seconds.*
B. Then, the flow of sub-events is to be presented – typically more or less in the order in which they happened. Adverbials support this order, for example *then, next, just a second later*, etc.
C. Deviation from this order must be explicitly marked, normally by adverbials such as *at the same time, just a bit earlier,* etc.

Communicative strategies of this type were already described in Ancient rhetoric. They have been more systematically studied in personal narratives (Labov 1972; Grimes 1975). But they are also necessary for other text types, whose internal temporal make-up is much less clear, for example to instructions of use, route directions, or simply descriptions. They also exhibit a complex temporal structure, albeit of a different type.

One way to look at how time is encoded in these different verbal tasks is to understand the whole text as an answer to an – explicit or implicit – question, the *Quaestio* (von Stutterheim and Klein 1989; von Stutterheim 1997). Faced with the same knowledge about a cosplex set of facts, a speaker may approach the task to describe them and their temporal relationships in very different ways. To a large extent, this depends on what the underlying quaestio is: *What happened to you? What did you observe? How do I bake a lemon cake? Where is the main station? What should I do to find a spouse?* The quaestio, whether implicit or explicitly asked by an interlocutor, imposes certain constraints on which parts of the speaker's underlying knowledge are to be put into words, how the information thus selected is packed into sentences, and how the sentences are arranged one after the other. The sentences that constitute the whole text are uttered (and heard or read) in a temporal order, and this temporal order can be used as a device to encode temporal properties of the "knowledge base". By knowledge base, I mean the speaker's activated knowledge of whatever he is asked to talk about by the quaestio. This can be his knowledge about an event that has he witnessed, such as a car accident, it can also be his knowledge about how to bake a cake, it can be his opinions on whether one should marry. In many cases, such a knowledge base offers a clear temporal structure, for example in the case of a car accident. Then, the speaker can linearise his text in accordance with this inherent temporal structure. This leads to a discourse strategy which is sometimes called "Principle of chronological order":

(20) PRINCIPLE OF CHRONOLOGICAL ORDER

Unless marked otherwise, the order in which the events are reported corresponds to the order in which they happened.

But this is only a special case. The knowledge which you activate when asked *Where is the station?* or *What does your living room look like?*, has no intrinsic temporal structure. In these cases, speakers normally "invent" a reasonable structure according to which they linearise the information; they may follow a gaze tour (Linde and Labov 1975; Ehrich and Koster 1983) or an imaginary wandering through the streets (Klein 1982). In a way, they "temporalise" static information. This is much more difficult in other cases, for example when you have to answer questions like *How does one play chess?* or *Why should one marry?* In each of these cases, subparts of the answer – the entire text – may involve temporal relations; but there is no overarching principle the speaker could follow when organising the text.

8. A new picture

There is a wealth of means to express various aspects of time; but even in the best-studied languages, these means, their form and their function, are only partly understood. In the preceding sections, I have tried to give an idea of the received picture, and I have pointed out a number of problems with this picture. In this concluding section, I will sketch some ideas about how one could go beyond the traditional views and thus possibly overcome at least some of the difficulties. They do not answer all of these problems; but they suggest a way in which they might be solved.

8.1. Clause-internal and clause-external temporal structures

Basic to this new picture is a sharp distinction between the bare *temporal structure* and the *descriptive properties* which linguistic expressions, words or constructions, associate with this temporal structure. Consider (21):

(21)   Eva seemed to have planned to mow the lawn at six.
        (t_0) t_1 t_2 t_3 t_4 t_5

Clearly, there is not just one "situation time" which is related to some other "clause-external time" t_0 but a whole set of time spans which are (a) temporally related to each other, and (b) characterised by certain descriptive properties. Very roughly, we have:

- t_0 is after t_1, the 'time of seeming'
- t_1 overlaps with t_2, the 'posttime of planning'
- t_2 is after t_3, the 'time of planning'
- t_3 is before t_4, the 'time of mowing'
- t_5 is most likely identical with t_4; but other readings are possible (see section 5.3, ex. 18)
- t_1 overlaps with t_2, and t_2 in turn overlaps with t_4.

Thus, a sufficiently general analysis of temporality must be able to account for quite complex temporal structures, rather than just relating the time of the situation and the moment of speech to each other. In particular, it must answer the following four questions:

A. Which time spans constitute the internal temporal structure of the clause?
B. How are these time spans related to each other?
C. Which **temporal** properties go with the various spans, i.e., properties such as position in relation to some other time span, duration and frequency?
D. Which **descriptive** properties go with the time spans? In (21), there is a time at which something seems to be the case, a time at which someone

apparently plans to do something, a time at which this planning is over, etc. This information is primarily provided by the descriptive content of the various verb forms; it can also stem from context.

Now, (21) is a relatively complex sentence. But essentially the same point can be made for sentences which are much simpler:

(22) Eva mowed the lawn.

Here, we have only one verb form in the past – *mowed*. It merges a finiteness marking (here *-ed*) and a non-finite component (*mow*), which, by virtue of its lexical meaning, provides certain temporal and descriptive properties. Very roughly, we have:

(23) 1. There must be a time t_1 at which Eva is "somehow active", for example swinging a sythe, pulling a lawn-mower, operating with some shears or whatever; the lexical verb leaves this to some extent open.[8]
2. There must be a time t_2, at which the grass is 'upright'.[9]
3. There must be a time t_3, at which the grass is 'on the ground'.
4. Various temporal relations obtain between these times. Thus, t_3 must be after t_2. The time t_1, the time at which Eva is active, must somehow overlap with t_2, i.e., the time at which the tree is upright; it may reach into t_3, but this is irrelevant for sentence 22 to be true.

These conditions do not exhaust the lexical content of the lexical verb *mow*. In particular, there is also a causal, and not just a temporal, connection between the whatever Eva does and and the fact that eventually, the grass is 'on the ground'. But this does not matter for present concerns.

Two conclusions may be drawn from this brief discussion. First, a clear distinction should be made between the 'temporal structure' itself (the temporal intervals and the temporal relations between them), on the one hand, and the descriptive information which goes with these intervals, on the other. The bare temporal structure would be exactly the same, if the sentence were *Eva left* or *Eva opened the window* – all that is different are the descriptive properties that are assigned to the various intervals. Second, temporal struc-

---

[8] The actitity could even consist in an instruction to someone else, as in *Louis IV built Versailles*.
[9] English does not provide us with good basic words for the first state and the second state of the grass; for the second state, we could perhaps best use the participle *mown* – which, of course, is derived.

ture as well as descriptive properties can be "packed" in different ways: they may be distributed over several words, they may also be found in a single verb form, here the word *mowed*.

Whatever this structure is and how it is expressed – it is not related to some time span outside the clause. As was discussed in section 2.3, the temporal anchor of a sentence need not be the moment of speech, and sometimes, it is not just one time span but a more complex structure. This invites the following picture. The expression of time in natural languages relates a CLAUSE-INTERNAL TEMPORAL STRUCTURE to a CLAUSE-EXTERNAL TEMPORAL STRUCTURE. The latter may shrink to a single interval, for example the time at which the sentence is utterared; but this is just a special case. The clause-internal temporal structure may also be very simple – it may be reduced to a single interval without any further differentiation, the "time of the situation"; but if this ever happens, it is only a borderline case. As a rule, the clause-internal temporal structure is much more complex, although the expression may be very simple, as (22) illustrates.

8.2. A few examples

If we want to understand how time is encoded in natural languages, we must first look at how a clause-internal temporal structure is built up in a particular language. The starting point of such an analysis is the lexical content of the verb (or a verb-like construction). This content can already have a rich internal temporal structure, as the example *mow* has illustrated. It is then enriched by all sorts of morphologic and syntactic operations, which yield various more complex forms (for example those in (21)). These operations render the temporal structure more and more complicated, they may also add new descriptive content. Let me illustrate this with a simple example, the (non-finite) verb form *be mowing* (as in [Eva be mowing the lawn]) which, very roughly, can be analysed in three steps:

(24) a. The lexical unit *mow* involves two distinct time spans, a "source time" – this is a time span at which Eva does something and the grass is "not on the ground" – a "target time" at which Eva is no longer active and the grass is "on the ground".[10]

---

[10] In Klein 1994, I have called these "source state" and "target state", respectively. This terminology (which was also used in sections 4 and 5 above) is suggestive but also somewhat unfortunate insofar as the notion of "state" merges bare time and descriptive properties.

b. The morphological operation which yields the form *mowing* adds a third time span which is a subinterval of the source time – it is within the time at which Eva is active and the grass is "not on the ground"; let me call this *ing*-time.
c. The syntactic operation which adds *be* and thus yields the new form *be moving* does not add a new time nor new descriptive properties; it just leads to a different syntactic construction which is then accessible to further operations; in particular, the topmost element can be made finite.

The resulting expression *be moving* already exhibits a rich temporal structure with accompanying descriptive properties. It is not yet a full sentence; the argument slots must be filled (yielding non-finite [Eva be mowing the lawn]), and it must be made finite. This is done by the appropriate morphological marking of the topmost verbal form, here *be*. In this way, it is related to the clause-external structure, in the simplest case the moment of speech. Various optional elements can be added, for example the temporal adverbial *at six*. The result is the sentence:

(25) Eva was mowing the lawn at six.

By uttering this sentence, the speaker asserts that at some time in the past, Eva was in the "source time" of some activity in whose "target time" the grass is "on the ground". His assertion is confined to this time; at this "topic time", the properties associated with the target time are not yet the case – the "situation is not completed but on-going". This explains the imperfective flavour of (25). Note, however, that this on-goingness is relative to the time about which the assertion is made – it is related to some time in the past.

Rather than making *be moving* finite, the process of forming complex structures might go on:

(24) d. The morphological operation which turns *be* into *been* and thus yields the form *been moving* adds a posttime to the *ing*-time; since the ing-time is a subinterval of the source time (the time at which Eva does something), it could still be the case that at this posttime, Eva is still active – but it could also be that her activity is over; it is only required that the new time is after an interval with the source time properties.
e. The syntactic operation which adds *have* and thus yields the new form *have been moving* does not add a new time nor new descrip-

tive properties; it just leads to a different syntactic construction which is then accessible to further operations; in particular, the topmost element can be made finite.

When this form is made finite, variables are filled and (optionally) an adverbial is added, we get:

(26) Eva had been mowing the lawn at six.

This says that at the time talked about, Eva had the "posttime properties" of being in the source time of mowing the lawn. It does not imply that at that time, she had finished mowing the lawn (= properties of target time reached). So, in a way, we have an "imperfective" within the perfect. What the perfect does, is therefore to assign "posttime properties"; what these properties are, depends on essentially on the meaning content of the underlying lexical verb and potentially other operations that have been applied to this verb.

Rather than making *have been moving* finite, we still could go on to form more complex verbal expressions, for example *plan to have been mowing*; this would add a pretime to the time of *have been mowing*, and this pretime is characterised as a "planning time". From there, we could proceed *to have planned to have been moving* or to *seem to have planned to have been mowing*, thus creating increasingly complex clause-internal structure. I will not elaborate on this here.

8.3. Classical notions re-defined

The brief exposition above is, of course, very crude. In particular, I believe that the temporal structure must be relativised with respect to the arguments of the verbs. In telic verbs such as *to open, to kill, to mow,* we must distinguish between time spans that are relevant for the subject ("x be active") and time spans that are relevant for the object ("y be not open – y be open", or "y be alive – y be dead"); similarly, the English auxiliary *have* has a slightly more complicated function than assumed in (24). The leads to more refined "time-argument structures" (see Klein (2002): but it does not affect the general idea.

I believe that an analysis along these lines will allow us to reconstruct classical notions such as tense, aspect and Aktionsart in a systematic and precise way, while keeping the intuitions that underly these notions. Tense, for example, serves to hook up the topmost verbal element of the clause-internal temporal structure to the clause-external temporal structure. In simple cases, there is only one verbal element, and the clause-external structure is the time of utterance, and this is the constellation which give rise to the classical notion of tense. The notion of grammatical aspect reflects some constellations *within* the clause-internal temporal structure, in particular those which are linked to the time of the top-most verbal element; the time of this element (the "topic time") is, so to speak, the joint between clause-external and clause-internal temporal structure. Aktionsarten are the inherent time-argument structures of verbs or of more complex expressions. This also accounts for the affinity between "grammatical aspect" and "lexical aspect"; they are special cases of the clause-internal temporal structure.

The exact way in which this functions in particular languages varies, of course, with the inherent temporal structure of lexical items, on the one hand, and of the morphosyntactic operations that can be used to form more complex expressions. Therefore, "achievements" in language A are often not like "achievements" in language B, and "imperfective" in language X is often not "imperfective" in language Y. Under the approach sketched here, we may not only be able to give a more precise account of classical notions such as tense and aspect and how they function in particular languages but also a broader picture of the many ways in which time can be encoded in human languages.

Acknowledgements

Many thanks to Leah Roberts who corrected my English.

References

Alexiadou, Artemis, Monika Rathert and Arnim von Stechow (eds.)
    2003   *Perfect explorations*. Berlin/New York: Mouton de Gruyter.
Bennett, Michel and Barbara H. Partee
    1978   *Toward the logic of tense and aspect in English*. Bloomington: Indiana University Linguistic Club.
Binnick, Robert
    1991   *Time and the Verb*. Oxford: Oxford University Press.
Bybee, Joan L., Revere Perkins and William Pagliuca
    1994   *The Evolution of Grammar: Tense, Aspect, and Modality in the Languages of the World*. Chicago: University of Chicago Press.

Comrie, Bernard
    1976   *Aspect. An Introduction to the Study of Verbal Aspect and Related Problems.* Cambridge: Cambridge University Press.
    1985   *Tense.* Cambridge: Cambridge University Press.
    1986   Tense in indirect discourse. *Folia linguistica* 20: 265–296.
Dahl, Östen
    1985   *Tense and Aspect Systems.* Oxford: Blackwell.
Dahl, Östen (ed.)
    2000   *Tense and aspect in the languages of Europe.* Berlin/New York: Mouton de Gruyter.
Declerck, Renaat, Susan Reed and Bert Cappelle
    2006   *The Grammar of the English Tense System.* Berlin/New York: Mouton de Gruyter.
Denison, David
    1993   *English historical syntax.* London/New York. Longman.
Dickey, Stephen M.
    2001   *Parameters of Slavic aspect.* Stanford: CSLI.
Dowty, David
    1976   *Montague grammar and word meaning.* Dordrecht. Reidel.
Ebert, Karen and Fernando Zuniga (eds.)
    2001   *Aktionsart and Aspectotemporality in Non-european languages.* Zürich: ASAS.
Ehrich, Veronika
    1992   *Hier und Jetzt. Studien zur lokalen und temporalen Deixis im Deutschen.* Tübingen: Niemeyer.
Ehrich, Veronika and Charlotte Koster
    1983   Discourse organisation and sentence form: The structure of room descriptions in Dutch. *Discourse Processes* 6: 169–195.
Elsness, Johan
    1997   *The perfect and the preterite in contemporary and earlier English.* Berlin/New York: Mouton de Gruyter.
Fenn, Peter
    1987   *A Semantic and Pragmatic Examination of the English Perfect.* Tübingen: Narr.
Forsyth, John
    1970   *A Grammar of Aspect,* Cambridge: Cambridge University Press.
Fabricius-Hansen, Cathrine
    2001   Wi(e)der and Again(st). In *Audiatur Vox Sapientiae*, Caroline Fery and Wolfgang Sternefeld (eds.), 100–130. Berlin: Akademie-Verlag.
Garey, Howard B.
    1957   Verbal Aspect in French. *Language* 33: 91–110.
Giorgi, Alessandra and Fabio Pianesi
    1997   *Tense and Aspect: From Semantics to Morphosyntax.* Oxford: Oxford University Press.

Grimes, Joseph
 1975 *The thread of discourse*. Mouton: The Hague.
Hamburger, Käte
 1956 *Die Logik der Dichtung*. Stuttgart: Klett.
Hamann, Cornelia
 1987 The Awesome Seeds of Reference Time. In *Studies on Tensing in English 1*, A. Schopf (ed.), 27–69. Tübingen: Niemeyer.
Hennig, Mathilde
 2000 *Tempus und Temporalität in geschriebenen und gesprochenen Texten*. Tübingen: Niemeyer.
Klein, Wolfgang
 1982 Local deixis in route directions. In *Speech, Place, and Action: Studies in Deixis and Related Topics*, Robert Jarvella and Wolfgang Klein (eds.), 161–182. New York: Wiley.
 1992 The present perfect puzzle. *Language* 68: 525–552.
 1994 *Time in Language*. London: Routledge.
 2001 Time and again. In *Audiatur Vox Sapientiae*, Caroline Fery and Wolfgang Sternefeld (eds.), 267–286. Berlin: Akademie-Verlag.
 2002 The argument-time structure of recipient constructions in German. In *Issues in Formal German(ic) Typological Studies on West Germanic*, Werner Abraham and Jan-Wouter Zwart (eds.), 141–178. Amsterdam/Philadelphia: John Benjamins.
Klein, Wolfgang, Ping Li and Henriette Hendricks
 2000 *Linguistic Theory* 18: 723–770.
Kratzer, Angelika
 1978 *Semantik der Rede*. Kronberg: Scriptor.
Krifka, Manfred
 1989 *Nominalreferenz und Zeitkonstitution. Zur Semantik von Massentermen, Pluraltermen und Aspektklassen*. München: Fink.
Labov, William
 1972 The Transformation of Experience in Narrative Syntax. In *Language in the Inner City*, William Labov (ed.), 354–396. Philadelphia: University of Pennsylvania Press.
Linde, Charlotte and William Labov
 1975 Spatial networks as a site for the study of language and thought. *Language* 51: 924–939.
Musan, Renate
 2002 *The German Perfect*. Dordrecht: Kluwer.
Noreen, Adolf
 1923 *Einführung in die wissenschaftliche Betrachtung der Sprache*. Halle: Niemeyer.
Portner, Paul
 2003 The (temporal) semantics and the (modal) pragmatics of the English Perfect. *Linguistics and Philosophy* 26: 459–510.

Paul, Hermann
    1886   *Prinzipien der Sprachgeschichte*, 2nd edition. Halle: Niemeyer.
Reichenbach, Hans
    1947   *Elements of Symbolic Logic*. Berkeley: University of California Press.
Reiff, Christian Philippe
    1928/9   *Grammaire raisonnée de la langue russe*. St. Petersburg (translation of a Russian grammar by N. I. Grech).
Smith, Carlota
    1997   *The Parameter of Aspect*, 2nd edition. Dordrecht: Kluwer.
von Stechow, Arnim
    1996   The different readings of Wieder 'Again'. A structural account. *Journal of Semantics* 13: 87–138.
von Stutterheim, Christiane
    1997   *Einige Prinzipien des Textaufbaus*. Tübingen: Niemeyer.
von Stutterheim, Christiane and Wolfgang Klein
    1989   Referential Movement in Descriptive and Narrative Discourse. In *Language Processing in Social Context*, C. F. Graumann and R. Dietrich (eds.), 39–76. Amsterdam: North Holland.
Taylor, C. C. W.
    1965   States, Activities, and Performances 2. *Proceedings of the Aristotelian Society* 39: 85–102.
Thieroff, Rolf
    1992   *Das finite Verb im Deutschen. Modus – Tempus – Distanz*. Tübingen: Narr.
Vendler, Zeno
    1957   Verbs and Times. *The Philosophical Review* 66: 143–160.
Williams, Christopher
    2002   *Non-progressive and progressive aspect in English*. Bari: Schena.
Xiao, Richard and Tony McEnery
    2004   *Aspect in Mandarin Chinese*. Amsterdam: Benjamins.

Temporal anaphora in a tenseless language[*]

Jürgen Bohnemeyer

1. Introduction

This chapter presents a portrait of a language that arguably lacks absolute (i.e., deictic) and relative (i.e., anaphoric) tenses and temporal connectives with meanings comparable to those of English *after*, *before*, *until*, and *while*. The language is Yucatec Maya. "Tenselessness", the absence of tenses from the grammar of a language, has been documented for a number of languages.[1] Yucatec goes beyond tenselessness in its simultaneous lack of temporal connectives of the indicated kind. However, the emphasis in this chapter is on tenselessness and on the question how Yucatec speakers manage to communicate about time in the face of it.

It is assumed in this chapter that tenses express binary ordering relations between the time about which an utterance makes a statement, asks a question, or issues a command, etc. – the **topic time** of the utterance, following

---

[*] Sections 2 and 3 of this chapter present a summary of chapters 4–7 of Bohnemeyer (1998b) and (2002). The research presented in these works was fully funded by the Max Planck Society. The examples used in section 3 are mostly new, though, and the analysis of one of the aspect-mood markers of Yucatec – the remote future ("predictive" in Bohnemeyer 1998b, 2002) – has been revised in light of new evidence. The account of temporal anaphora in section 4 is an informal version of the analysis I presented at the SULA 5 conference in São Paulo in May 2007 (though using different material for illustration). This analysis is an update of the one in Bohnemeyer (1998b, 2000a/b, 2002), preserving its Gricean core, but attempting a simpler, more concise, and more rigorous formulation (and integration into the DRT framework, which is not discussed in the present paper).

[1] Cf., e.g., Bittner (2005, 2007) and Shaer (2003) on Kalaallisut (or West Greenlandic); Comrie (1976: 82–84) on Igbo and Yoruba; Comrie (1985: 50–53) on Burmese and Dyirbal; and Li & Thompson (1981: 184, 213–215) on Mandarin. Tenselessness may also be discussed in terms of the absence of tense marking in particular utterances, rather than in the entire grammar. For instance, Smith, Perkins & Fernald (2007) argue that Navajo has a future-non-future tense system, but that tense marking is optional in this language. The interpretation of tenseless utterances in Navajo, based on these authors' observations, appears to rely on principles similar to those proposed for Yucatec in section 4.

Klein 1994 (see also chapter 2 of this book) – and, in the case of deictic tense, the time *at* which the utterance is made or interpreted (the **coding time**), or, in the case of anaphoric tenses and temporal connectives, some time mentioned in discourse – the **reference point**.[2] In this framework, Yucatec can be characterized as a language in which the topic times of utterances are not constrained vis-à-vis utterance times or reference points by the morphosyntactic form of the clause (but adverbials may of course be used to determine them). This claim is defended in section 3 and possible exceptions are discussed there as well. Beyond making the case for the existence of languages such as Yucatec, the main concern of this chapter is the question of how Yucatec speakers determine the topic times of utterances in the absence of explicit coding. The proposal developed in section 4 is that they rely on inference mechanisms of **temporal anaphora** which are shared with Indo-European languages and, presumably, universal.

Temporal anaphora is the contextual determination of topic times. Temporal anaphora resolution depends on many factors, including the semantics of aspectual and modal operators (aside from lexical semantic properties, rhetorical structure, and world knowledge). The main emphasis in the present chapter is on the impact of aspectual operators. Consider (1):

(1) a. Floyd entered. Sally made a phone call.
    b. Floyd entered. Sally was making a phone call.

According to the standard analysis of temporal anaphora in Discourse Representation Theory (DRT; cf. Kamp 1979; Kamp & Rohrer 1983; Kamp & Reyle 1993; Hinrichs 1981, 1986), the second sentence in (1a) introduces a new topic time (in the present terminology) following that of the first sen-

---

[2] In the case of assertions, the topic time is the (implicitly or explicitly given) time for which it is claimed that some state of affairs holds. If the utterance concerns an event, the topic time may be different from the time of the event; the ordering relation between topic time and event time (or "situation time", in Klein's terminology) is expressed by viewpoint-aspectual operators. For example, imperfective viewpoints place the topic time inside the time of the event, while perfective viewpoints inversely place the time of the event inside the topic time. Klein's "topic time" corresponds broadly to the adaptation of Reichenbach's (1947) "reference point" in the DRT literature. In contrast, in the present framework, "reference point" is used for the time interval with respect to which topic times may be determined. A pure anaphoric tense on this account is an operator that expresses a relation between topic time and some reference point in the same way a pure deictic tense expresses a relation between topic time and coding time.

tence, with the event time of the phone call included in the new topic time, whereas the progressive in (1b) introduces a state to the discourse representation whose run time includes the topic time, and that reference time is unchanged from the first sentence, as if the second sentence tracked it anaphorically.[3]

The account sketched in this chapter presents the determination of topic time vis-à-vis coding time in Yucatec as a special case of temporal anaphora. The existence of temporal anaphora in a tenseless language such as Yucatec is itself not surprising, but nevertheless remarkable: it shows that temporal anaphora it is not an anaphoric meaning component of tense morphemes, as assumed in Partee (1973) and the DRT literature. Topic times play a role in the interpretation of utterances whether or not these are tensed, and the principles involved in their contextual resolution are the same in tensed and tenseless languages.

The account of temporal anaphora developed here treats the inferences involved in topic time resolution as Gricean generalized conversational implicatures. Part of the evidence in favor of this approach is the non-monotonicity of the inferences. Contra Bittner (2008), the inferences are just as defeasible in the "aspectually fully explicit" Yucatec as they are in English. The Gricean analysis accounts for both the non-monotonicity of the inferences, which is attributed to vagueness in the DRT literature, and for their default character. The Gricean account offers a parsimonious alternative to the assumption of special principles governing the deictic interpretation of tenseless sentences, as proposed in Smith, Perkins & Fernald (2007).

The following section presents a thumbnail sketch of the resources involved in temporal reference in Yucatec. Section 3 summarizes the evidence for tenselessness, and section 4 lays out the analysis of temporal anaphora.

2. A sketch of the Yucatec grammar and lexicon of temporal reference

2.1. Background

Yucatec is the largest member of the Yucatecan branch of the Mayan language family. It is spoken by 759,000 people in the Mexican states of Campeche, Quintana Roo, and Yucatán, and approximately 5,000 people in the

---

[3] Hence the term "temporal anaphora" originally coined in Partee (1973) and identified as a metaphor in Partee (1984).

Cayo District of Belize (*Ethnologue*, Gordon 2005).[4] Yucatec is the language whose autodenomination, *Maya*, has been adapted by scholars to name the Mayan language family.

Yucatec is a polysynthetic language in the sense that syntactic relations tend to have morphological reflexes at the word level and in the sense that a single content word, in combination with the necessary function words and inflections, may – and frequently does – constitute a clause. In terms of the morphological complexity of content words, Yucatec is situated towards the high end among Mayan languages, but in a more central position among Mesoamerican languages overall. Yucatec is mostly head-initial, and in particular verb-initial, but this fact is somewhat obscured by the high frequency of left-dislocations and focus constructions in discourse. The language has a typologically unusual argument marking system in which the single core argument of an intransitive clause patterns with either the actor or the undergoer argument of an active transitive clause depending on aspect-mood inflection.[5]

There are five sources of overtly expressed temporal information in Yucatec: lexical items, the verb inflection system, adverbials, connectives, and certain nominal suffixes. The following subsections address these in turn.

2.2. Lexical items

Yucatec has a complex system of lexical categories not all niceties of which are as yet fully understood. The following thumbnail sketch summarizes the analysis presented in Bohnemeyer (2002, 2004) and Bohnemeyer & Brown 2007. This analysis proposes a taxonomic organization with a top-level split between verbs and other categories. Verbs obligatorily inflect for a functional category called **status** (following Kaufman 1990) in all syntactic environments in which they function as verbs. Status inflection expresses, in a single suffix position, distinctions of viewpoint aspect, modality, and

---

[4] Based on 2005 Census data, which register a decline by more than 40,000 speakers age five or older since 2000 (http://www.inegi.gob.mx/est/contenidos/espanol/rutinas/ept.asp?t=mlen10&c=3337).

[5] This is an atypical "split-S" or "active-inactive" pattern in which the morphological treatment of the core argument of intransitive clauses is fully determined, not by the lexical semantic properties of the verb, but instead by inflectional distinctions akin to those found in split-ergative case marking systems; cf. Bohnemeyer 2004 and references therein.

illocution; cf. below. The architecture of the status system differs from language to language; in Yucatec, there are five subcategories: completive, incompletive, subjunctive, extra-focal, and imperative status. Every verb form must be morphologically specified for exactly one of these subcategories in any syntactic environment – there is no finiteness contrast with regard to status inflection. In contrast, stative predicates are incompatible with status inflection. All content words – other than verb roots or stems – can form stative predicates by themselves, without the need of a copula.

(2) Túumben   le=nah=o'
    new(B3SG) DET=house=D2[6]
    'The house is/was/will be new'

Stative predicates include nouns, adjectives, and deverbal statives. Verb stems fall into five different "conjugation" classes distinguished by paradigms of allomorphs of the status suffixes, which are termed "active", "inactive", "inchoative", "positional", and "transitive" in Bohnemeyer (2002, 2004).[7] As shown in Bohnemeyer (2001, 2002, 2004), membership in these verb stem classes is strikingly well motivated in terms of causativity and

---

[6] Abbreviations in morpheme glosses: 1/2/3 – 1st/2nd/3rd person; A – cross-reference set-A (clitics; possessor/highest-ranking argument); ACAUS – anticausative derivation/middle voice; ALT – conditional/disjunctive/polar-question-focus particle; APP – applicative derivation; ASS – assurative modal (promises, offers, commitments); ATP – antipassive derivation; B – cross-reference set-B (suffixes; theme/lowest-ranking argument); CAL – past instance of calendrical interval; CAUS – causative derivation; CAUSE – causal preposition; CL – (numeral/possessive) classifier; CMP – completive status; CON – connective; D1/2/3/4 – proximal/distal-anaphoric/text-deictic/negation-locative clause-final particle; DES – desiderative modal; DET – determiner; EXTRAFOC – extra-focal status; F – feminine/endearment; HS – hearsay; IMM – immediate past; IMPF – imperfective aspect; IN – inanimate (classifier); INC – incompletive status; INCH – inchoative derivation; IRR – irrealis mood; NEC – necessitive modal (circumstantial/conditional necessity); NEG – negation; OBL – obligative modal (social obligations, schedules, etc.); PASS – passive voice; PEN – penative modal (counter-factual); PL – plural; PREP – generic preposition; PROG – progressive aspect; PROSP – prospective aspect; PROX – proximate future; PRV – perfective aspect; REC – recent past; REL – relational nominal derivation; REMF – remote future; REMP – remote past; REP – repetitive/reversative; SG – singular; SR – subordinator; SUBJ – subjunctive status; TERM – terminative (= perfect-like) aspect ; TOP – topic (= left-dislocation).

[7] V denotes a vowel whose quality is copied from the preceding root vowel.

the distinction between processes and state changes. Both processes and state changes are dynamic eventualities that involve change in some property over time. Process descriptions either leave the event participant to whom the changing property is attributed unexpressed (e.g., *paint*, as opposed to *paint a picture/the wall*) or fail to specify a reference point with respect to which the change could be evaluated (e.g., *walk*, as opposed to *walk to the station*); cf. Bohnemeyer & Swift 2006. The Yucatec active class includes underived process verb stems – especially manner of motion (e.g., *áalkab* 'to run'; *bàab* 'to swim') and sound emission verbs (e.g., *hàayab* 'to yawn; òok'ol* 'to cry'), denominal verb stems (e.g., *e'l* 'egg', 'testicle' > 'to ovulate', 'to lay eggs'; *míis* 'broom' > 'to sweep'), antipassive stems derived from transitive roots (e.g., *k'ay* 'to sing sth. (song)' > *k'àay* 'to sing'; *tus* 'to lie to sb.' > *tùus* 'to lie'), and all intransitive stems derived from Spanish roots, regardless of their semantics – the most important exception to motivation in terms of process semantics (e.g., *áatrasáar* 'to become late'; *duràar* 'to last'; *gáanar* 'to win'). The majority of process stems describe "internally caused" activities (including all of the examples cited above); but externally caused events are likewise lexicalized in active roots (though without encoding their cause; e.g., *balak'* 'roll', *chíik* 'shake', 'rattle'; *péek* 'move', 'wiggle') (cf. Smith 1978 and Levin & Rappaport-Hovav 1995 on the distinction between internal and external causation).

Table 1. Status polysemy and verb stem classes

STATUS STEM CLASS	INCOM-PLETIVE	COMPLETIVE	SUB-JUNCTIVE	EXTRA-FOCAL	IMPERATIVE
active	-Ø	-nah	-nak	-nahik	-nen
inactive	-Vl	-Ø	-Vk	-ik	-en
inchoative	-tal	-chah	-chahak	-chahik	N/A
positional	-tal	-lah	-l(ah)ak	-lahik	-len
transitive active	-ik	-ah	-Ø / -eh	-ahil	-Ø / -eh
passive	\'/...-Vl / -a'l	\'/...-ab / -a'b	\'/...-Vk / -a'k	\'/...-ik / -a'bik	N/A

Inactive, inchoative, and positional stems describe different kinds of state changes without expressing their causes. Inchoative stems are derived from adjectives and, less regularly, nouns. Positionals – or "dispositionals" (Bohnemeyer & Brown 2007) – lexicalize non-inherent spatial properties that may be thought of as "manners" of location. Distinctions that enter the conceptualization of dispositions include "position", or more generally support/suspension (e.g., *kul* 'sit', *wa'l* 'stand', *chil* 'lie', *xol* 'kneel', *nak* 'lean', *choh* 'hang', *hoch'* 'droop'); blockage of motion (e.g., *tak'* 'be stuck to something', *kap* 'be stuck between two things'); orientation in the gravitational field (e.g., *haw* 'lie face up', *nok* 'lie face down', *tsel* 'lie on side', *ch'eb* 'be tilted', 'lean to one side'); and configurations of parts of an object with respect to each other (e.g., *hen* 'be sprawled', *nik* 'be scattered', *hay* 'be spread out', *much'* 'be in a pile', *tsol* 'be lined up in a row', *sop'* 'be coiled up'). Dispositional roots require derivational morphology for the formation of both stative and intransitive dynamic stems, the latter expressing the "assume position" sense (Levin 1993). However, a majority of dispositional roots produces transitive verb stems without overt derivation. Finally, inactive roots include "path verb" roots (*bin* 'go', *tàal* 'come', *u'l* 'return', *máan* 'pass', etc.); "phase" (= aspectual) roots (e.g., *ho'p'* 'begin', *ts'o'k* 'end'); and also a few roots referring to bodily or cognitive state changes (e.g., *síih* 'be born', *kim* 'die', *ah* 'wake up', *wèen* 'fall asleep') and creation/destruction type events. The inactive class is fed by the anticausative derivation, and passivized transitive stems have a status pattern closely related to that of inactive stems (see Table 1).[8]

Transitive roots overwhelmingly express caused state changes (e.g., *kach* 'break', *xot* 'cut'), but a few also lexicalize forced contact (e.g. *hats'* 'hit', *koh* 'beat', *yet'* 'massage') and changes in mental states in which an "experiencer" outranks the theme or stimulus (e.g., *il* 'see', *na't* 'guess', 'intuit', 'reason', 'understand', and *kan* 'learn'). All intransitive verb roots produce derived transitive stems via the causative and "applicative" derivations.

There are very few, if any,[9] stative verbs; meanings lexicalized in stative verbs in English, such as 'know', 'love', or 'have', are expressed by relational nouns and elements of other categories. Thus, the parts-of-speech system of Yucatec seems to be semantically motivated in terms of dynamicity more clearly than Indo-European languages.

Membership in the active class as opposed to the other verb stem classes is determined along the lines of the process-change distinction, not in terms of telicity, along the lines of the distinction between Vendlerian activities vs. accomplishments and achievements (as suggested by Lucy 1994). Thus,

---

[8] The anticausative is treated as a derivation in Bohnemeyer 2004, but is argued to have also voice-like traits in Bohnemeyer 2007.

[9] It is not clear that there are *any* verbs in the language that do not have at least one dynamic sense.

the inactive and inchoative verb stem classes include many "degree achievements" projecting atelic event descriptions unless some "degree of change" is overtly specified (Abusch 1985; Bertinetto & Squartini 1995; Dowty 1979: 88–91; Hay, Kennedy, & Levin 1999). This is illustrated in (3) for the inchoative *kàabal-tal* 'lower' (derived from the adjective *kàabal* 'low'):

(3) Le=mùunyal=o' ts'u=chúun-ul   u=kàabal-tal,
     DET=cloud=D2 TERM:A3=start\ACAUS-INC A3=low-INCH.INC
     'The cloud, it had started going down (lit. lowering),'

   káa=h-tàal   le=ìik'=o',
   CON=PRV-come(B3SG) DET=wind=D2
   '(when/and then) the wind came'

   káa=t-u=pul-ah   le=mùunyal=o'.
   CON=PRV-A3=throw-CMP(B3SG) DET=could=D2
   '(when/and then) it drove (lit. threw) the cloude away.'

   Ts'u=kàabal-tal   le=mùunyal=o'?
   TERM:A3=low-INCH.INC DET=cloud=D2
   'Had the cloud gone down?'

   -Ts'o'k ka'ch u=kàabal-tal   káa=h-pu'l-ih.
   TERM formerly A3=low-INCH.INC CON=PRV-throw\PASS-CMB(B3SG)
   'It had gone down (when/and then) it was driven (lit. thrown) away.'

In (3) the speaker proposes a scenario to a native speaker consultant in which a cloud starts sinking before it is pushed away by the wind. He asks the consultant whether it is possible in this context to say that the cloud had already become lower when it was pushed away. The consultant answers in the affirmative.

Conversely, many active intransitive stems have salient telic "performance object" interpretations – especially antipassive stems such as *k'àay* 'sing' in (4):

(4) Pedro=e'   táan u=k'àay,
     Pedro=TOP PROG A3=sing\ATP
     'Pedro, he was singing,'

   káa=t-u=k'at-ah   u=báah Pablo.
   CON=PRV-A3=cross-CMP(B3SG) A3=self Pablo
   '(when/and then) Pablo interfered.'

Pedro=e'   t-u=p'at-ah   u=k'àay.
Pedro=TOP PRV-A3=leave-CMP(B3SG) A3=sing\ATP
'Pedro', he stopped singing.'

Be'òora=a' ts'o'k=wáah   u=k'àay   Pedro?
now=D2   TERM=ALT   A3=sing\ATP Pedro
'Now, has Pedro sung?'

- Ma'=h=bèey-chah   u=k'àay=i'.
NEG=PRV=thus-PROC.CMP(B.3.SG) A3=sing\ATP=D4
'He didn't manage to sing (lit. his singing didn't become possible).'

In this case, the question asked of the consultant is whether somebody who is singing, but interrupted in the course of it, can be said to have sung. The consultant denies this, assuming that the singer in the scenario wasn't able to complete his song.

(3)–(4) illustrate the application of the entailment pattern known as the "imperfective paradox" (Dowty 1979: 133) in telicity tests. There are no syntactic correlates of telicity in Yucatec. There is no distinction between duration (or *for*-type) and time-frame (or *in*-type) adverbials (cf., e.g., Krifka 1992); the same verbs are used in translations of "She spent X-time VERB-ing" and "It took her X-time to VERB"; and there are no "aspectualizers" or "phase verbs" that only occur with either telic or atelic verbal cores (cf. Freed 1979).[10] It is not clear, at present, why telicity is reflected in the syntax of some languages but not in that of others.

2.3. The verb inflection system

In main clauses, the verb inflection system specifies two positions in which aspectual and modal information is obligatorily expressed: status inflection and the pre-verbal aspect-mood (AM) markers. Table 2 summarizes the distribution and semantics of the five status categories.

---

[10] A verbal core is a maximal projection of a verb that dominates all of the verb's arguments and obliques, like a subject-internal verb phrase. The term is borrowed from Van Valin & LaPolla 1997. There is no evidence of a (subject-external) verb phrase in Yucatec. Verbal cores combine with aspect-mood markers to constitute clauses in Yucatec (see below); clauses and verbal cores are thus in complementary distribution.

Table 2. Distribution and semantics of the five status categories

Category	Distribution	Semantics
completive	independent verbal cores w/ perfective AM marker	perfective aspect and "assertive" modality
incompletive	dependent verbal cores; independent verbal cores w/ imperfective AM marker	imperfective aspect and "assertive" modality
subjunctive	dependent verbal cores; optative clauses; irrealis subordinate clauses	perfective aspect and "non-assertive" modality
extra-focal	manner focus construction (dependent verbal core)	perfective aspect and "assertive" modality
imperative	imperative sentences	perfective aspect and "directive" illocution

Affirmative declarative main clauses are constituted by the combination of a verbal core with exactly one of the 15 AM markers. Two of these – the perfective and imperfective AM – are prefixes; the verbal cores they combine with are independent. The remaining 13 AM markers are stative predicates which take a dependent verbal core as their sole argument.[11] Each AM marker assigns a status category marked on the verb; Table 3 below lists the combinations.[12]

---

[11] There is one exception: the prospective AM marker *mukah*. *Mukah* takes an oblique dependent verbal core and carries a set-B suffix cross-referencing the actor argument if the latter is transitive and the single argument if it is intransitive. Status marking on the verb likewise depends on the verb's transitivity (see Table 2), as it does in certain control constructions with lexical matrix predicates (see Bohnemeyer in press).

[12] Table 3 provides only the base forms of the AM markers; they frequently form portmanteaus with the set-A clitics of the verbal core. Necessitive *k'a'náan* has an apparent synonym *k'abéet*; some speakers use the two interchangeably, while others prefer *k'a'náan*. Other, apparently more obsolete, general necessity modals occur as well. The distinction between AM markers and lexical stative predicate that combine with dependent verbal cores is not sharply delimited. For instance, Ayres & Pfeiler (1997) and Vapnarsky (1999) treat *sùuk* 'custom', 'habit' as a habitual marker, while Bohnemeyer (2002) considers *sùuk* a lexical stative complement-taking predicate.

The semantics of the preverbal AM markers is briefly discussed below and in section 3. Relative and "topic" clauses (see section 2.5), focus constructions,[13] and negation involve alternative AM systems. Details vary from construction to construction (cf. Bohnemeyer 2002: 116–129); but all of these constructions are governed by what may be characterized as a realis-irrealis mood contrast, where realis mood includes reference to individual present or past events, the irrealis future time reference, and habitual and generic reference may be subsumed under either category depending on the construction. In the realis mood, depending on the construction, either the system in Table 3 or a simplified version involving bare incompletive and subjunctive forms without preverbal AM markers are used. The irrealis is expressed using a bare incompletive form under negation and the irrealis subordinator *kéen* plus subjunctive status (in some cases also incompletive, but only with intransitive verbs) in the other constructions. The examples in (5) illustrate the realis-irrealis contrast for content questions targeting the undergoer of a transitive verb. The realis example (5a) has the perfective aspect marker with past time reference; the irrealis example (5b) the irrealis subordinator *kéen* and subjunctive status on the verb with future time reference:

(5) a. Ba'x   **t**-a=mèet-**ah**?
        what(B3SG) **PRV**-A2=do:APP-**CMP**(B3SG)
        'What did you do?', 'What is it that you did?'

     b. Ba'x   **kéen**   a=mèet-eh?
        what(B3SG) **SR.IRR** A2=do:APP-**SUBJ**(B3SG)
        'What will you do/are you going to do?', 'What is it that you will do/are going to do?'

---

[13] There is evidence suggesting that all focus constructions, including all content questions, are clefts in Yucatec and that all clefts use relative clause constructions as nominal predicates; cf. Bricker (1979); Bohnemeyer (2002: 116–129); Tonhauser (2003). For an alternative viewpoint, cf. Gutiérrez Bravo (2006).

Table 3. Preverbal AM markers and status assignment

Subset of AM markers	AM marker	Status category
AM prefixes	perfective h-(V_{itr})/t- (V_{tr})	completive
	imperfective k-	incompletive
Aspectual AM predicates	progressive *táan*	incompletive
	terminative *ts'o'k*	incompletive
	prospective *mukah*	incompletive (V_{itr}) / subjunctive (V_{tr})
Modal AM predicates	obligative *yan*	incompletive
	necessitive *k'a'náan*	incompletive
	desiderative *táak*	incompletive
	assurative *he'* ...=*e'*	incompletive
	penative *òolak*	subjunctive
Metrical AM predicates	remote future *bíin*	subjunctive
	remote past *úuch*	subjunctive
	recent past *sáam*	subjunctive
	immediate past *táantik* ...=*e'*	incompletive
	proximate future *ta'itak*	incompletive

Finally, dependent verbal cores also occur as arguments or obliques of lexical predicates and as complements of prepositions. In these constructions, status marking is controlled by the head (cf. Bohnemeyer 2002: 91–101 for details).

It is conceivable that the presence vs. absence of an AM marker in a clause or phrase is correlated with – and thus an expression of – the semantic property of finiteness. Finiteness can be understood as the property of a clause or verbal projection of requiring a particular "topic component" for its interpretation (Klein 2006). In languages that distinguish finite from non-finite verb forms, projections of finite verb forms must be understood as making an assertion, asking a question, etc., about a particular topic time, a particular topic place, (a) particular possible world(s), a particular set of entities, etc. In contrast, projections of nonfinite verb forms are interpreted, not with respect to their own topic components, but as part of some superordinate syntactic structure which is ultimately "hooked up" to a topic component via a finite verb form. Yucatec lacks a distinction between finite and non-finite verb forms. All verb forms are marked for status, and all status categories except for the completive occur both in independent and dependent verbal cores. It seems plausible that, in first approximation, (dependent or independent) verbal cores are inherently non-finite and that it is the combination with a preverbal AM marker that maps the "sentence base" expressed by the verbal core into a topic component. Future research will have to examine how finiteness is expressed – if at all – in focus constructions, etc., and under negation.

The semantics of the five status categories is analyzed in Bohnemeyer (2002: 216–242), with the upshot represented in Table 2. The perfective-imperfective distinction is conceptualized here following Klein 1994, i.e., in terms of a temporal relation between the time of the eventuality described in an utterance and the "topic time" for which the denotation of the utterance is computed. Perfective aspect places the event time inside the topic time; imperfective aspect conversely places the topic time inside the event time. Subjunctive status, like completive, is perfective, i.e., encodes inclusion of the event time of the verbal core in the topic time. The subjunctive differs from the completive in that it "decouples" the interpretation of the sentence or clause formed by combining the verbal core with an AM marker or a lexical matrix predicate from the realization of the event described by the verbal core ("non-assertive" modality). The subjunctive occurs as an optative mood with matrix predicates of desire. It is likewise triggered by the "penative" AM marker (from Latin *paene* "almost"), which has a counterfactual semantics and is used to describe events that almost, but not quite, become reality. The remote future and recent and remote past markers, which likewise govern the subjunctive, cardinally quantify over the distance between topic time and event time. None of these entails **event realization** (cf. Bohnemeyer & Swift 2004). The immediate, recent, and remote past markers presuppose, rather than to assert, realization of the event described by the verbal core. Hence, when they are negated, it is to deny the recency/remoteness of the event described by the verbal core – not that the event happened:

(6) Ma' sáam sùunak le=kòombi=o';…
 NEG REC turn\ATP:SUBJ(B3SG) DET=van=D2
 'It's not a while ago that the bus returned;…'

 a. …inw=a'l-ik=e', h-ts'o'k mèedya òora.
  A1SG=say-INC(B3SG)=TOP PRV-end(B3SG) half hour
  '…I think it was half an hour ago.'[14]

---

[14] The recent past marker is a "same day" past marker, i.e., is used for events that happened earlier on the day that includes topic time. By preemption, it receives a non-immediacy implicature due to its contrast with the immediate past marker. This implicature is interpreted in (6b) as metalinguistically negated.

b. ??...tuméen ma' sùunak=i'.
CAUSE   NEG turn\ATP:SUBJ(B3SG)=D4
'...because it hasn't returned yet.'

Presupposition may also motivate the occurrence of the bare subjunctive in certain focus constructions (including the so-called "agent-focus" form) with perfective reference. The irrealis subordinator *kéen* illustrated above likewise assigns subjunctive status.

"Extra-focal" is a status category exclusively reserved to a particular kind of focus constructions: manner focus constructions, in which the focused element refers to a property of the event described by the "extra-focal" subordinate clause. If that clause is perfective, its verb is in the extra-focal. Example (7) illustrates:

(7)   Domìingo-ak=e'   ma'+lo'b   h-hàats'-nahik-en.
      Sunday-CAL=TOP NEG+bad PRV-hit\ATP-EXTRAFOC-B1SG
      'Last Sunday, well is how I batted.'

Imperative forms are used exclusively in commands. They do not combine with preverbal AM markers, and the set-A prefixes cross-referencing the addressee as the single argument of intransitive imperatives and the A-argument of transitive imperatives is deleted.

Turning to the AM markers, the semantics of the five aspectual AM markers is sketched in Figure 1. The analysis presupposes the framework of Klein 1994, with one modification: prospective and perfect are understood as placing topic time in the run time of some pre- or post-state causally related to the target event, rather than merely in *times* preceding and following the event, respectively. The imperfective AM marker has progressive interpretations mostly in focus constructions. Outside focus constructions, it occurs in irrealis contexts, i.e., in clauses with habitual, generic, and future time reference whose subordinate counterparts trigger irrealis marking in the appropriate syntactic environments. Only the progressive reading is represented in Figure 1. The perfective AM marker has both perfective interpretations proper and (resultative) perfect interpretations; Bohnemeyer (2002: 246–250) argues that this is a case of vagueness. The question whether the perfective marker incorporates a past tense component is discussed in section 3.

Four of the five modal AM markers involve universal quantification over the possible worlds in "modal bases" (Kratzer 1977, 1981, 1991) of different kinds. The obligative is used to express that the sentence base eventuality is realized in all worlds consistent with the social obligations of the referent of the S/A argument, with schedules, and with the ordinary course of events in general. The desiderative AM marker indicates that the base eventuality is realized in all possible worlds in which the (in particular, bodily) desires of the referent of the S/A argument are met. The assurative AM is used for universal quantification over a modal base in which promises, offers, and commitments by the referent of the S/A argument are fulfilled; in these uses, it conveys a certain illocutionary flavor. But the assurative is also used with epistemic modal bases. In this case, it expresses certainty; but, perhaps due to the contrast with the obligative and prospective markers – both of which are used for reference to events expected to occur in "inertia worlds" (Dowty 1979; Landman 1992; Portner 1998), based on evidence of causally related pre-states that obtain in the topic situation – epistemic uses of the assurative tend to have a strong flavor of subjective opinion or estimate. The fourth universal-force modal, the necessitive, is used to express circumstantial or conditional necessity – the idea that something must happen if something else is to be allowed to happen or to be avoided.[15]

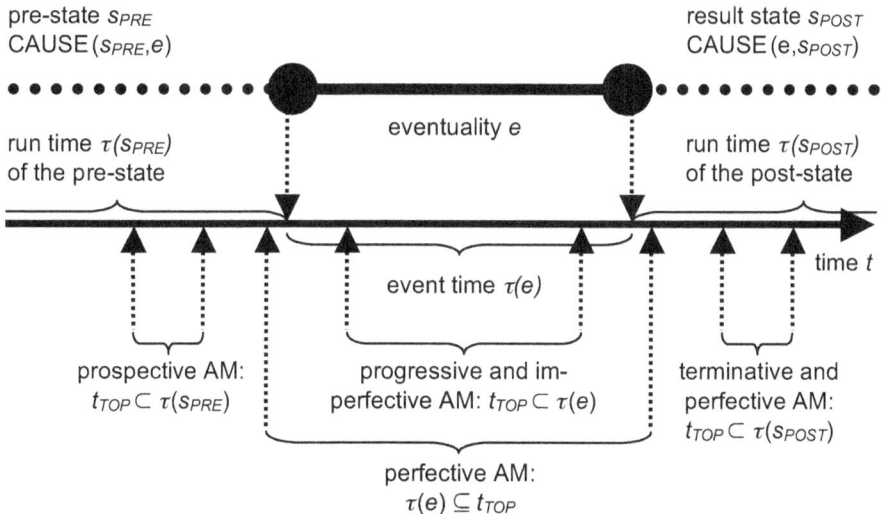

Figure 1. Semantics of the aspectual AM markers (key: CAUSE(a,b) – event a is/was/will be the/a cause of event b; S_{PRE} – pre-state; S_{POST} – result-state; $\tau(a)$ – run time of a; t_{TOP} – topic time)

---

[15] While there are thus four universal-force modals integrated into the paradigm of AM markers, existential modality – i.e., possibility – is expressed by inflected lexical predicates. The reason for this curious asymmetry is unknown.

The "penative" AM marker *òolak* is used to indicate that an instance of the base event was at some past topic time about to happen, but subsequently did not – a purely counterfactual sense of "almost". It can be viewed as a counterfactual mirror image of the prospective and proximate-future markers. Its grouping with the four universal-force modals is obviously somewhat arbitrary; it does not perfectly fit into any of the subcategories of AM markers distinguished here.

The five metrical AM predicates quantify the distance between the run time of the base eventuality and topic time. The denotation of proximity in the proximate future marker (how soon is "soon"?) is purely a matter of contextual standards. The rare remote future marker seems to be primarily used to convey negation of this subjective sense of proximity. Consultants consistently volunteer the information that the remote future suggests that the speaker has no idea when, if at all, the base event will happen.[16] The three metrical past markers form a scale[17] the central member of which, the recent past marker *sáam*, prototypically extends into the relative past up to the boundary between the day that includes topic time and the preceding day, i.e., functions as a "hodiernal past" marker.[18] By preemption, the immediate and remote past markers are interpreted to the exclusion of the recent past marker. In the case of the immediate past marker *táantik ...=e'*, this has the effect that the latter is used predominantly for events that happened at a distance which is short compared to the lengths of a day (no more than a few hours, and often less than an hour), while the remote past marker *úuch* is used at a distance considered long by the same standard (typically several days ago).

---

[16] The remote future marker was mischaracterized as a "predictive" modal marker in Bohnemeyer (1998b, 2000a, 2002, 2003). The research on which this body of work was based did not produce evidence of the marker. The conjecture of modal semantics took off from the observation that the marker is associated with the genre of prophecies (Vapnarsky 1995). However, recent elicitation has shown it to be part of the "degrees-of-remoteness" system. See section 3 for examples.

[17] This is not an entailment scale in the sense of Horn 1972 since the three terms denote distances from topic time and thus entail upper bounds. The markers' interpretation is nonetheless subject to pragmatic enrichment, as mentioned below.

[18] This cut-off point is blurred by the frequent use of *sáam* to express that something happened earlier than expected (by the speaker and/or the addressee), even if it happened within the domain of the immediate or remote past. See also section 3.

The five metrical distance predicates differ in several ways from "metrical tense" phenomena described in the literature (e.g., Comrie 1985: ch.4; Dahl 1984, 1985: 120–128). It is argued in section 3 that instead of constraining possible topic times of the utterance vis-à-vis coding time or some reference point, as true (metrical) tenses would, they express the distance between topic time and event time as a state that obtains at topic time.

2.4. Adverbials

Within the domain of temporality, adverbials are used to express calendrical or clock time indices or the event times of eventualities. There is a handful of lexical indexical adverbs, including terms locating days with respect to coding time (e.g., *ka'ho'lheak* 'the day before yesterday', *sáamal* 'tomorrow') and the "topic time shifters" *be'òora ...=a'/=e'* 'now' and *ka'ach(il)* 'formerly'. The latter are briefly discussed in section 3. Apart from these, temporal adverbials are mostly headed by spatial prepositions such as *ich(il)* 'in', 'inside' and relational nouns such as *táan* 'front'. These adverbials specify time intervals, but do not encode temporal ordering relations between these times and the topic or event time of the utterance. Thus, *ich(il ti') ts'e'ets'ek k'ìin* may be used to translate 'in a few days' (referring to an event in the (relative) future), '((with)in a few days' (referring to a time span within which some event was or will be completed), 'for a few days' (referring to the duration of a process or interval), or 'a few days ago', depending on aspect-mood marking.

(8) Pwes to'n=e', **ich ts'e'ts'ek k'ìin** hóok'-ok-o'n.
    well PREP:B1PL=TOP **in a.few sun** exit-SUBJ-B1PL
    'Well, as for us, it was a few days ago that we left.'

(9) Pwes to'n=e', **ich ts'e'ts'ek k'ìin** kéen hóok'-ok-o'n.
    well PREP:B1PL=TOP **in a.few sun** SR.IRR exit-SUBJ-B1PL
    'Well, as for us, it is in a few days that we will leave.'

Bohnemeyer (in press) suggests that this may be a reflex of the literal spatial meanings of the heads, which encode what Jackendoff 1983 calls "place functions", but no locative or path relations. For instance, *ich le=nah=o'*, formed with the same preposition *ich*, translates 'in the house', 'into the house', 'out of the house', and 'through the house', depending on the verb the phrase combines with.

2.5. Connectives

As in other languages, subordinate clauses may serve to introduce the time of the event they describe as a point of reference in discourse. There are, however, no connectives in Yucatec that encode ordering relations between time intervals such as 'after', 'before', or 'while'. There is likewise no word for 'when', although there are a few idioms that serve as temporal anaphors without expressing ordering relations. An example is the phrase *chéen ya'lo'* 'at that time', literally 'it will say that' in (10), using a form of the irrealis subordinator *kéen*:

(10) Chéen ka'=sùunak-ech   t-u=láak'   ha'b=e',
     SR:IRR REP=turn\ATP:SUBJ-B2SG PREP-A3=other year=TOP

   chéen   y=a'l-∅=e',
   SR:IRR A3=say-SUBJ(B3SG)=TOP

   táantik in=mèet-ik   le=nah=o'.
   IMM   A1SG=DO:APP-INC(B3SG)   DET=house=D2

   '(When) you return here next year, at that time (lit. (when) it will say that), I will have just build the house.'

The first clause in (10), *chéen ka' sùunakech tuláak' ha'be'* '(when) you return next year', introduces a reference point taken up first by the phrase that functions as a temporal anaphor (itself a clause) – which is actually redundant in this example – and then by the main clause, whose topic time it is understood to mark. These subordinate clauses determine the topic time for the main clause much like temporal clauses in Indo-European languages. However, syntactically, they are not adverbials, but share the morphological and ordering properties of left-dislocated arguments – they are "topic clauses" (Bohnemeyer 1999c). Neither the clause that introduces the reference time nor the phrase *chéen ya'lo'* are morphologically "flagged" for expressing temporal reference points. They are marked as topics by their position and the clause-final particle that follows them (the topic marker =*e'* in (10)); but in an appropriate context, the same clauses could be interpreted as headless relative clauses. This ambiguity is illustrated in (11):

(11) **Le=k-u=tàas-a'l=e',**   k-u=bo'l-t-a'l.
     DET=IMPF-A3=come:CAUS-PASS.INC=TOP IMPF-A3=pay-APP-PASS.INC
     '(What) is brought is paid for / (when) it is brought, it is paid for.'
     (based on Blair & Vermont Salas 1965–67 11.1.25)

Instead of connectives expressing ordering relations, both "topic clauses" of the kind illustrated in (10)–(11) and main clauses frequently carry what may be described as "aspectual connectives" (Bohnemeyer 1998b: 485–503). An example is *káa*, which occurs exclusively with the perfective AM marker. The perfective AM marker has both perfective and result-state interpretations (e.g., the same clause can be used to convey 'The car broke down' and 'The car is broken down', not unlike the simple past in English). *Káa* forces the perfective interpretation. Result-state reference, like all notional viewpoint aspects except for the perfective, requires a contextual reference point (a "natural temporal reference point", see section 4) to determine its topic time. *Káa* blocks interpretation of the clause with respect to a tracked topic time. Put differently, it signals "resetting" of the topic time variable. This device is used above all in narrative discourse. The result can be a sequential or simultaneous interpretation depending on context. In (12), the preferred interpretation is sequential. The order of events is inferred from the order of clauses on the basis of stereotype implicatures in this case (see section 4). In (13), the preferred interpretation is simultaneous for most speakers because the two actions reported have different agents. This makes it clear that *káa* does not express an ordering relation.

(12) Pedro=e'   **káa**=t-u=ts'íib-t-ah   hun-p'éel
    Pedro=TOP CON=PRV-A3=write-APP-CMP(B3SG) one-CL.IN
    kàarta=e', **káa**=t-u=ts'u'uts'-ah   hun-p'éel chamal
    letter=TOP CON=PRV-A3=suck-CMP(B3SG)   one-CL.IN cigarette
    'Pedro, (when) he wrote a letter, he smoked a cigarette'
    (preferred interpretation sequential)

(13) **Káa**=t-u=ts'íib-t-ah   hun-p'éel kàarta Pedro=e',
    CON=PRV-A3=write-APP-CMP(B3SG)   one-CL.IN letter   Pedro=TOP
    Juán=e',   **káa**=t-u=ts'u'uts'-ah   hun-p'éel chamal
    Juán=TOP CON=PRV-A3=suck-CMP(B3SG) one-CL.IN cigarette
    '(When) Pedro wrote a letter, Juán smoked a cigarette'
    (preferred interpretation for most consultants simultaneous)

In (12)–(13), the topic clauses refer to the times of the events they describe. The topic times of the main clauses are interpreted with respect to these event times, though not necessarily as coinciding with them, since the main clauses likewise contain the "topic time reset" connective *káa*. The results are the observed interpretations. It is also possible for topic clauses to refer to times before or after the event they describe. Aspectual and modal expressions are used for this purpose. Example (14) shows a topic clause that sets a topic time for the main clause preceding the topic clause event, the speaker's arrival:

(14) Ma' k'uch-uk-en=e', káa=h-hóok' leti'
     NEG arrive-SUBJ-B1SG=TOP CON=PRV-exit(B3SG) DET:PREP
     '(When) I had not yet arrived, (and) she left'

The topic clause in this case uses negation and subjunctive status to refer to a time prior to the realization of the topic clause event. This construction is the closest Yucatec equivalent of the English connective *before*.

This section has provided an overview of the various types of lexical and morphosyntactic resources involved in the expression of temporality in Yucatec. The following sections summarize the evidence for tenselessness (section 3) and offer an account of temporal anaphora in Yucatec (section 4).

3. Tenselessness

Yucatec is a tenseless language in the sense that the morphosyntactic form of the clause does not constrain its use with topic times in the present, past, or future of coding time (absolute = deictic tense) or some other reference point (relative = anaphoric tense). "Morphosyntactic form" is taken here to mean syntactic structure plus/including inflection. To demonstrate tenselessness, it needs to be shown that any Yucatec clause regardless of syntactic structure and inflection is compatible with both topic times in the present, past, and future of coding time and topic times in the present, past, and future of some other reference point. The discussion in the present chapter focuses on clauses that can constitute sentences by themselves ("main" or "independent" clauses). Subordinate clauses are briefly taken up at the end of this section. The syntactic variation that needs to be taken into account when evaluating the claim of tenselessness for main/independent clauses is primarily the selection of the preverbal "aspect-mood" (AM) marker. As mentioned in the previous section, the "status" suffix on the verb is determined by the AM marker. The examples in (15) illustrate the 15 AM markers.

(15) a. *Perfective AM – completive status*
     T-in=mèet-**ah**   le=nah=o'
     PRV-A1SG=do:APP-**CMP**(B3SG) DET=house=D2
     'I built the house'

b. *Imperfective AM – incompletive status*
   K-in=mèet-**ik**   le=nah=o'
   IMPF-A1SG=do:APP-**INC**(B3SG) DET=house=D2
   'I (would) build the house'

c. *Terminative AM – incompletive status*
   Ts'o'k in=mèet-**ik**   le=nah=o'
   TERM   A1SG=do:APP-**INC**(B3SG) DET=house=D2
   'I (will) have/had built the house'

d. *Progressive AM – incompletive status*
   Táan in=mèet-**ik**   le=nah=o'
   PROG A1SG=do:APP-**INC**(B3SG) DET=house=D2
   'I am/was/will be building the house'

e. *Prospective AM – subjunctive status on transitive verbs, incompletive on intransitive verbs*
   Mukah in=mèet-∅   le=nah=o'
   PROSP A1SG=do:APP-**SUBJ**(B3SG) DET=house=D2
   'I am/was/will be going to build the house'

f. *Necessitive AM – incompletive status*
   K'a'náan in=mèet-**ik**   le=nah=o'
   NEC   A1SG=do:APP-**INC**(B3SG) DET=house=D2
   'I must/had to/will have to build the house'

g. *Obligative AM – incompletive status*
   Yan in=mèet-**ik**   le=nah=o'
   OBL A1SG=do:APP-**INC**(B3SG) DET=house=D2
   'I (will) have/had to build the house'

h. *Assurative AM – incompletive status*
   He' in=mèet-**ik**   le=naho'
   ASS A1SG=do:APP-**INC**(B3SG) DET=house=D2
   'I will/would indeed (agree to) build the house (you shall see)'

i. *Desiderative AM – incompletive status*
   Táak in=mèet-**ik**   le=nah=o'
   DES   A1SG=do:APP-**INC**(B3SG) DET=house=D2
   'I (will) want(ed) to build the house'

j. *Penative AM – subjunctive status*
   Òolak in=mèet-∅   le=nah=o'
   PEN   A1SG=do:APP-**SUBJ**(B3SG) DET=house=D2
   'I (will have/had) almost built the house'

k. *Remote future AM – subjunctive status*
   Bíin   in=mèet-Ø   le=nah=o'
   REMF   A1SG=do:APP-SUBJ(B3SG)   DET=house=D2
   'It is/was/will be a long time before I build the house'

l. *Proximate future AM – incompletive status*
   Ta'itak   in=mèet-ik   le=nah=o'
   PROX   A1SG=do:APP-INC(B3SG)   DET=house=D2
   'I will/would soon build the house'

m. *Immediate past AM – incompletive status*
   Táantik   in=mèet-ik   le=nah=o'
   IMM   A1SG=do:APP-INC(B3SG)   DET=house=D2
   'I (had/will have) just built the house'

n. *Recent past AM – subjunctive status*
   Sáam   in=mèet-Ø   le=nah=o'
   REC   A1SG=do:APP-SUBJ(B3SG)   DET=house=D2
   'I (had/will have) built the house already/not long ago'

o. *Remote past AM – subjunctive status*
   Úuch   in=mèet-Ø   le=nah=o'
   REMP   A1SG=do:APP-SUBJ(B3SG)   DET=house=D2
   'I (had/will have) built the house long ago'

The 15 examples can be used, with the exceptions to be discussed below, in each of the following three contexts in the position marked "_____". Each context determines the topic time of the utterance following it in the position marked "_____". In (16), Jorge is asking Pedro whether his has built the house he had been planning to build. This question introduces a time frame starting with Jorge's last visit two years earlier and leading up to the present of the utterance. In (17), Jorge asks whether Pedro will build the house in the future, and Pedro begins his response by shifting the topic time to Jorge's next visit a year into the future. In the final context, (18), Jorge asks whether Pedro's house is new, and Pedro sets the topic time of his response to the time of Jorge's last visit two years earlier.

(16) *Diagnostic context: topic time leading up to utterance time*

   Jorge'   táantik   u=k'uch-ul   x-Yaxley=e'.
   Jorge:TOP   IMM   A3=arrive-INC   F-Yaxley=D3
   'Jorge, he has just arrived in (the village of) Yaxley.'

H-ts'o'k   ka'-péel   ha'b káa=h-sùunah
PRV-end(B3SG) two-CL.IN year CON=PRV-turn:CMP(B3SG)
'It has been two years since he returned...'

t-u=kàah-al=o'.
PREP-A3=reside\ATP-REL=D2
'...to his country.'

T-uy=ohel-t-ah
PRV-A3=knowledge-APP-CMP(B3SG)
'He knew...'

táak u=mèet-ik   u=nah-il   Pedro.
DES A3=do:APP-INC(B3SG) A3=house-REL Pedro
'...that Pedro wanted to build a (lit. his) house.'

Ba'x=e'   ma' t-uy=ohel-t-ah
what=TOP   NEG PRV-A3=knowledge-APP-CMP(B3SG)
'But he didn't know...'

wáah h-bèey-chah-ih.
ALT   PRV-thus-INCH.CMP-CMP(B3SG)
'...whether he was able to do it (lit. whether it became possible).'

Káa=t-uy=il-ah   Pedro te=kàaye=o',
CON=PRV-A3=see-CMP(B3SG)   Pedro PREP:DET=street=D2
'He saw Pedro in the street,'

káa=t-u=k'áat+chi'-t-ah.
CON=PRV-A3=wish+mouth-APP-CMP(B3SG)
'and he asked him.'

Káa=t-u=núuk-ah   Pedro=e':
CON=PRV-A3=answer-CMP(B3SG) Pedro=TOP
'Pedro answered:'

"_____"

(17) *Diagnostic context: topic time in the future of utterance time*

Jorge'   ta'itak u=sùut   t-u=kàah-al=o'.
Jorge:TOP IMM   A3=turn\ATP PREP-A3=reside\ATP-REL=D2
'Jorge, he would soon return to his country.'

T-uy=ohel-t-ah
PRV-A3=knowledge-APP-CMP(B3SG)
'He knew...'

táak u=mèet-ik u=nah-il Pedro.
DES A3=do:APP-INC(B3SG) A3=house-REL Pedro
'...that Pedro wanted to build a (lit. his) house.'

Ba'x=e' ma' t-uy=ohel-t-ah
what=TOP NEG PRV-A3=knowledge-APP-CMP(B3SG)
'But he didn't know...'

wáah yan u=bèey-tal.
ALT OBL A3=thus-INCH.INC
'...whether he would be able to do it (lit. whether it would become possible).'

Káa=t-uy=il-ah Pedro te=kàaye=o',
CON=PRV-A3=see-CMP(B3SG) Pedro PREP:DET=street=D2
'He saw Pedro in the street,'

káa=t-u=k'áat+chi'-t-ah.
CON=PRV-A3=wish+mouth-APP-CMP(B3SG)
'and he asked him.'

Káa=t-u=núuk-ah Pedro=e':
CON=PRV-A3=answer-CMP(B3SG) Pedro=TOP
'Pedro answered:'

"Chéen ka'=sùunak-ech t-u=láak' ha'b=e',
SR:IRR REP=turn\ATP:SUBJ-B2SG PREP-A3=other year=TOP
'When you return next (lit. the other) year,...'
___"

(18) *Diagnostic context: topic time in the past of utterance time*

Jorge', t-uy=ohel-t-ah
Jorge=TOP PRV-A3=knowledge-APP-CMP(B3SG)
'Jorge, he learned...'

t-u=mèet-ah u=nah-il Pedro.
PRV-A3=do:APP-CMP(B3SG) A3=house-REL Pedro
'...that Pedro had built a (lit. his) house.'

káa=t-u=k'áat+chi'-t-ah
CON=PRV-A3=wish+mouth-APP-CMP(B3SG)
'He asked him...'

wáah túumben le=nah=o'.
ALT new(B3SG) DET=house=D2
'...whether the house was new.'

Káa=t-u=núuk-ah Pedro=e':
CON=PRV-A3=answer-CMP(B3SG) Pedro=TOP
'Pedro answered:'

"Káa=h-tàal-ech way h-ts'o'k ka'=p'éel ha'b=e',
CON=PRV-come-B2SG here PRV-end(B3SG) two=CL.IN year=D3
'When you came here two years ago,…'
_____."

Most of the utterances in (15) are readily interpretable in these three contexts. For illustration, (19) has the remote past AM marker with future time reference and (20) the remote future AM marker with past time reference:

(19) Chéen ka'=sùunak-ech t-u=láak' ha'b=e',
SR:IRR REP=turn\ATP:SUBJ-B2SG PREP-A3=other year=TOP
úuch in=mèet-Ø le=nah=o'
REMP A1SG=do:APP-**SUBJ**(B3SG) DET=house=D2
'When you return next (lit. the other) year, I will have built the house long ago'

(20) Káa=h-tàal-ech way h-ts'o'k ka'=p'éel ha'b=e',
CON=PRV-come-B2SG here PRV-end(B3SG) two=CL.IN year=D3
bíin in=mèet-Ø le=nah=o'
REMF A1SG=do:APP-**SUBJ**(B3SG) DET=house=D2
'When you came here two years ago, it was going to be a long time before I would build the house'

However, there are a number of principled exceptions. First of all, there are pragmatic issues that may affect the naturalness of various markers in these contexts.[19] Secondly, the imperfective aspect marker (15b) has a progressive

---

[19] It is of course difficult to find a single example sentence format such that all 15 markers are equally natural in this format in all three contexts. The recent past marker (15n) is considered infelicitous by some consultants in all three contexts because it is commonly used to indicate that the target event happened/was realized earlier than expected/hoped/feared etc. This presupposition makes no sense in the contexts in (16)–(18). Most of the modal markers are more likely to be accepted with topic times shifted into the past or future of utterance time when occurring in the complement of a verb of cognition or a speech act verb or the like. The perfective AM marker is used for both perfective and result state reference. In the narrative context (18), the connective *káa* is preferred to be added at the left edge of (15a) to force the perfective interpretation.

meaning in focus constructions, but outside those is compatible with all and only those topic times that in subordinate clauses trigger irrealis marking (see section 2.3): habitual, generic, and future topic times. It is thus anomalous in descriptions of the building of an individual house in present (16) and past contexts (18), but is interpretable in the future time reference context of (17). Finally, and most significantly for present purposes, the perfective (15a) and "penative" (counterfactual) AM markers (15j) are incompatible with future time reference. The most straightforward explanation of this restriction would be that the markers have past (or non-future; the perfective is compatible with blow-by-blow-reporting-style contexts) meaning components. However, the perfective AM marker is in fact used with future time reference in conditional antecedents:

(21) a. Wáah **t**-a=ts'o'k-s-**ah** le=nah
ALT **PRV**-A2=end-CAUS-**CMP**(B3SG) DET=house

te=mèes k-u=tàal=o',
PREP:DET=month IMPF-A3=come=D2

hi'n=bo'l-t-ik tèech be'òora=a'.
ASS:A1SG=pay-APP-INC(B3SG) PREP:B2SG now=D1

'If you build the house next month, I will pay you now.'

b. Wáah **káa**=ts'o'k-s-Ø le=nah
ALT **SR**:A2 =end-CAUS-**SUBJ**(B3SG) DET=house

te=mèes k-u=tàal=o',
PREP:DET=month IMPF-A3=come=D2

hi'n=bo'l-t-ik tèech be'òora=a'.
ASS:A1SG=pay-APP-INC(B3SG) PREP:B2SG now=D1

'If you build the house next month, I will pay you now.'

In (21), a hypothetical conditional is used to offer the addressee a deal. For this purpose, either the perfective (a) or the subordinator *káa* in combination with subjunctive status (b) can be used without discernible semantic difference. Notice that the perfective does not convey relative past time reference in (21) either: the reference point for an anaphoric-tense interpretation would have to be the event time of the main clause, which however lies in the relative past, not future, of the topic time of the conditional clause in (21). If the semantic contribution of the perfective in conditionals such as (21a) is compositional, then the perfective cannot have a (deictic or anaphoric) past tense semantics.

The perfective AM marker occurs with future time reference in conditional antecedents, but not in any clause that asserts, questions, or presupposes propositions. I have argued (Bohnemeyer 1998b, 2000a, 2002) that this is due to a combination of the fact that the perfective marker entails event realization (Bohnemeyer & Swift 2004), i.e., factuality, with a principle that bars treating the realization of future events as fact:

(22) **Modal commitment constraint (MCC)**
The realization of events in the (relative or absolute) future cannot be asserted, denied, questioned, or presupposed as fact. Assertions, questions, and presuppositions regarding the future realization of events, or the failure thereof, require specification of a modal attitude on the part of the speaker.

The MCC requires clauses used to assert, deny, question, or presuppose realization of future events to express a modal attitude towards the realization of the event: necessity, desire, agreement, prediction, etc. Since the modal markers that express these attitudes cannot co-occur with the perfective AM marker in the same clause, the perfective is excluded from the relevant contexts by the MCC. The MCC is a language-specific constraint. However, similar principles have been reported for other tenseless languages (Comrie 1985: 50–53 for Burmese and Dyirbal; Bittner 2005 for Kalaallisut/West Greenlandic).

The MCC accounts the unavailability of the "penative", i.e., counterfactual, AM marker with future time reference since this marker negates realization of the target event. But why does the MCC not exclude the terminative aspect marker and the immediate, recent, and remote past markers from occurring with future time reference? All of these markers entail realization of the event described by the verbal core. However, these markers have stative meanings: they serve to assert, deny, or question, not the realization of the target event, but the result state of the target event (terminative AM) or the state of the target event's immediacy (immediate past AM), recency (recent past AM), or remoteness (remote past AM). Stativity of these markers is evident from their incompatibility with event time adverbials. This is illustrated for the terminative AM marker in (23). The adverbial *ho'lheak* 'yesterday' can only be interpreted as a topic time adverbial in (23a), rendering the sentence both pragmatically (the speaker is asking whether the addressee was during the day before the utterance in the state of having met their brother) and syntactically odd (topic time adverbials are preferred to be left-dislocated). To obtain the event time interpretation (i.e., to ask whether the addressee met the speaker's brother the day before the utterance), the perfective AM marker is used, as in (23b):

(23) a. ?Ts'o'k aw=il–ik   in=suku'n   ho'lheak?
        TERM   A2=see-CMP(B3SG)   A1SG=older.brother   yesterday
        'Had you met my brother yesterday?'

   b. T-aw=il-ah   in=suku'n   ho'lheak, he'bix
      PRV-A2=see-CMP(B3SG)   A1SG=older.brother   yesterday like
      t-a=tukul-ah=e'?
      PRV-A2=think-CMP(B3SG)=D3
      'Did you meet my brother yesterday, as you had planned?'

Incompatibility with event time adverbials can be established in the same fashion for the immediate, recent, and remote past AM markers. This incompatibility suggests that it is only the result state of the event in the case of the terminative marker and the state of the event having occurred at a certain distance from topic time in the case of the metrical tense markers that is accessible to adverbial modification. And it is the stativity of the markers in question that explains why they are not barred from future topic times by the MCC. All stative predicates of Yucatec occur freely with arbitrary topic times – the MCC does not apply to state descriptions:

(24) (Káa=h-tàal-ech   way   h-ts'o'k   ka'=p'éel   ha'b=e',
     CON=PRV-come-B2SG   here PRV-end(B3SG)   two=CL.IN   year=D3
     /chéen ka'=sùunak-ech   t-u=láak'   ha'b=e',)
     SR:IRR REP=turn\ATP:SUBJ-B2SG   PREP-A3=other   year=TOP
     túumben   le=nah=o'
     new(B3SG)   DET=house=D2
     '(When you came here two years ago/when you return next (lit. the other) year,) the house is/was/will be new'

The rationale behind this dichotomy in the grammatical treatment of propositions that concern the realization of events and propositions about states deserves further attention.

The case against an anaphoric-tense analysis of the Yucatec AM markers rests on Occam's Razor: the semantic contribution of a given marker in a given utterance can be either relative tense or some modal or aspectual meaning, but should not be assumed to be both except in the face of compelling evidence. Given this principle, the relative tense analysis of a marker is defeated by demonstrating aspectual or modal meaning components. Such meaning components are established on the basis of failure to entail event realization (this applies to all modal AM markers, the imperfective, progressive, and prospective aspect markers, and the remote and immediate future markers) or incompatibility with event time adverbials (this applies to the terminative aspect marker and the five AM markers expressing degrees of remoteness). The examples in (25)–(26) illustrate failure to entail event realization for the progressive (25) and obligative (26) AM markers:

(25) Káa=h-tàal-ech   way h-ts'o'k   ka'=p'éel ha'b=e',
     CON=PRV-come-B2SG here PRV-end(B3SG) two=CL.IN year=D3
     táan in=mèet-**ik**   le=nah=o'.
     PROG A1SG=do:APP-**INC**(B3SG) DET=house=D2
     Ba'x=e'   ma' h-bèey=chah
     what=TOP NEG PRV-thus=INCH.CMP(B3SG)=D4
     in=ts'o'k-s-ik   tuméen h-k'oha'n-chah-en.
     A1SG=end-CAUS-INC(B3SG) CAUSE PRV-sick-INCH.CMP-B1SG
     'When you came here two years ago, I was building the house. But, I wasn't able to finish it because I became ill.'

(26) Káa=h-tàal-ech   way h-ts'o'k   ka'=p'éel ha'b=e',
     CON=PRV-come-B2SG here PRV-end(B3SG) two=CL.IN year=D3
     yan in=mèet-**ik**   ka'ch   le=nah=o'.
     OBL A1SG=do:APP-**INC**(B3SG) formerly DET=house=D2
     Ba'x=e'   ma' h-bèey=chah=i',
     what=TOP NEG PRV-thus=INCH.CMP(B3SG)=D4
     tuméen h-k'oha'n-chah-en.
     CAUSE PRV-sick-INCH.CMP-B1SG
     'When you came here two years ago, I had to build the house. But, it didn't work out because I became ill.'

Incompatibility with event time adverbials has been exemplified for the terminative AM marker in (23). The remote future marker has both diagnostic properties: it fails to entail event realization (27) and is incompatible with event time adverbials (28).

(27) **Bíin** in=mèet-Ø   le=nah=o',   ba'x=e',   ma'
     REMF A1SG=do:APP-**SUBJ**(B3SG) DET=house=D2 what=TOP NEG

inw=ohel wáah yan u=bèey-tal.
A1SG=knowledge(B3SG) ALT OBL A3=thus-INCH.INC
'It will be a long time before I build the house, but I don't know whether it will be possible.'

(28) ***Bíin** in=mèet-Ø le=nah
REMF A1SG=do:APP-**SUBJ**(B3SG) DET=house
te=àanyo k-u=tàal=o'.
PREP=year IMPF-A3=come=D2
intended: 'I will build the house next year.'

The AM markers of temporal distance deserve special attention in the context of the tenselessness analysis. As mentioned in section 2, these differ from better studied "metrical tense" systems (cf., e.g., Comrie 1985: ch.4; Dahl 1984, 1985: 120–128) in a number of respects. Most importantly, as demonstrated here, they do not encode absolute tense. Neither is their use obligatory for reference to an event at the specified distance from topic time – they are used to emphasize the degree of remoteness much the same way adverbials such as *recently* and *a long time ago* are in English. They could be analyzed as optional anaphoric/relative metrical tenses. However, as shown above, they are stative predicates that do not permit event time specifications. Instead of specifying the distance between topic time and coding time, as absolute metrical tenses would, or the distance between topic time and some reference point, as relative metrical tenses would, they specify the distance between topic time and the time of the event described by the verbal core. They are thus not tenses in the sense of the definition given in the beginning of this section: they do not constrain the topic time of the utterance vis-à-vis coding time or some other reference point. Instead, their semantics concerns the relation between topic time and event time, much like that of the aspectual and modal AM markers.

Subordinate clauses and verbal cores can be divided into three classes: (a) embedded verbal cores which occur as complements of matrix predicates or adpositions – these are interpreted with respect to the topic time of the matrix; (b) subordinate clauses which show the same aspect-mood marking system as independent clauses – these are thus subject to the same argumentation advanced above for independent clauses; and (c) subordinate clauses that show the reduced AM-marking systems briefly discussed in section 2.3. As mentioned, these reduced systems are all governed by a realis-irrealis mood contrast. In this case, the fact that the same irrealis form used for (absolute and relative) future time reference is also used for habitual and generic reference makes a tense analysis implausible from the start.

Yucatec does, of course, have means of explicitly constraining the topic time of an utterance. Adverbials and subordinate clauses as illustrated in most of the examples in this section will do the job just fine. This includes two adverbs whose semantics is similar if not identical to that of present and past tenses, respectively: the "topic time shifters" *be'òora ...=a'* 'now', which restricts topic time to overlap with coding time, and *ka'ch(il)* 'formerly', which situates topic time in the past of utterance time. The latter is illustrated in the second line of (26) above. These are not considered tenses here because they are clearly not required by the morphosyntactic form of the sentence – they are just adverbs that optionally occur in the same positions as a host of other adverbs. As a result, their pragmatics is also quite different from that of tense marking in Indo-European languages. However, it seems quite possible that *optional* tense markers in languages that have them function similarly to the Yucatec topic time shifters. *Be'òora ...=a'* 'now' is predominantly used to implicate that a state asserted to hold at topic time did not hold at some relevant earlier time, and *ka'ch(il)* triggers the inverse implicature that the state asserted to hold at topic time does not hold any longer at coding time. For example, in (26), *ka'ch(il)* is used to indicate that the state of "obligation" to build the house no longer holds at utterance time. Of course, tense marking in English carries the same implicatures; the difference is that the Yucatec topic time shifters are used *only* when this implicature is intended to be conveyed.

4. Temporal anaphora in Yucatec

Section 3 has summarized the evidence suggesting that Yucatec is a tenseless language. Tenses serve to constrain the topic time of an utterance to the present, past, or future of coding time (in the case of absolute tense) or some reference point (in the case of relative tense). It is safe to assume that it is as important for Yucatec speakers as it is for English speakers to be able to distinguish narrative accounts of past events from predictions of future events or declarations of intentions about future events and, for example, descriptions of habits and statements of general rules. The question in the present section is not whether Yucatec speakers are able to infer that the topic time of any given utterance lies in the present, past, or future of coding time or some reference point; the question is how exactly they do this.

A standard claim made in discussions of tenselessness is that adverbials can be used to compensate for the lack of tense markers. While certainly not false, this is misleading to the extent that it suggests that adverbials are more frequent or play a more important role in discourses of tenseless languages than in those of tensed ones. At least as far as Yucatec is concerned, this is not the case. The events narrated in Yucatec folk tales are not anymore anchored to a calendrical time scale than those of *Hansel and Gretel* (see below for an illustration), and if a Yucatec speaker wishes to convey that the bus I have been waiting for has already left or that a house is on fire in another part of the village or that they are planning to get married, they are perfectly able and in fact likely to do so without using any temporal adverbials.

The argument to be advanced in this section is that to determine the order between the topic time of an utterance and its coding time, Yucatec speakers rely on the same "mechanism" speakers of English and other better studied languages rely on to determine topic times in context. It so happens that this mechanism is not needed to determine the relation between topic times and coding times in tensed languages such as English because these relations are expressed by tense markers. However, the mechanism plays a key role in determining temporal relations between clauses in connected speech in tensed and tenseless languages alike. The mechanism in question is, of course, that of **temporal anaphora**. Examples of temporal anaphora have been presented throughout the previous sections. Two excerpts from larger texts may help getting a flavor for the matter. Example (29) is a passage from a demon story (discussed in detail in Bohnemeyer 2003).

(29) Le=òotsil máak=o', káa=h-bin te'l ich
DET=poor person=D2 CON=PRV-go(B3SG) there in
'The poor man, he went (out) there to'

le=kòol=o'. Ti', bin, yàan te=ka'nal=o',
DET=clear\ATP=D2 there HS EXIST(B3SG) PREP:DET=high=D2
'the *milpa* (swidden, lit. 'clearance'). There he was, they say, up high (i.e., in a tree),'

chéen káa=t-y=il-ah
only káa=PRV-A3=see-CMP(B3SG)
'(and) he saw'

u=tíip'-il, bin, le=ba'l=o'; túun tàal
A3=appear-INC HS DET=thing=D2 PROG:A3 come
'the thing (i.e., the demon) appear, they say; it was coming'

The first clause describes an event of a man going to work in his *milpa* or swidden. The next clause is stative and describes the man's location up in a tree. The topic time of this state description is inferred to follow that of the first clause (a point to be commented on below). The tree is understood to be either on the *milpa* or on the way there (we do not actually learn whether the man ever reached his goal, and the first clause does not entail this, contrary to its gloss). The third clause, like the first, is an event description in the perfective aspect, introduced by the connective *káa* briefly discussed in section 2, which is characteristic of narrative discourse. It introduces the man's perception of the approaching demon. The topic time of this clause is understood to be the same as that of the preceding state description: the man is up in the tree and sees the demon from there. The fourth and final clause features the progressive aspect marker. It refers to the demon's approach. Its topic time is understood to be the same as that of the preceding two clauses. The aspect markers stipulate that the perception event is included in this topic time, whereas the times of the man's being up in the tree and the demon's approach contain the topic time.

In the second excerpt, the speaker talks about the hurricane Roxana which hit his village in 1996 about a week before the recording.

(30) Káa=h-k'uch-o'n   túun   way te=kàah-il
      CON=PRV-arrive-B1PL so.then here PREP:DET=live-REL
      'So then (when) we arrived here in the village'

      x-Yaxley-il=e',   k=il-ik=e'   tuláakal
      F-Yaxley-REL=TOP   IMPF:A1PL=see-INC(B3SG)=TOP all
      'of Yaxley, we saw (that) all'

      máak=e',   táan uy=a'l-ik-o'b=e'
      person=TOP   PROG A3=say-INC(B3SG)-3PL=TOP
      'people (i.e., everybody), they were saying'

      hach   ts'-uy-u'b-ik-o'b   ti'   ràadyo=e'
      really TERM-A3=perceive-INC(B3SG)-3PL   PREP radio=TOP
      '(that) they had really heard on the radio'

      túun   tàal le=siklòon=o'
      PROG:A3   come DET=cyclone=D2
      '(that) the hurricane was coming'

The first clause is marked for perfective aspect and introduces the event of the speaker's arrival in the village. As a "topic clause" (see section 2.3, 2.5), it sets the topic time for the following main clause, which in fact is the topic time for the entire passage. The main clause contains the perception verb *il* 'see' marked for imperfective aspect. This combination is an idiom frequently used in first-person narratives to describe the narrator's realization of previously unknown facts. Semantically, this idiom is interpreted perfectively. In (30), the realization is understood to take place at or after the speaker's arrival. The object of the speaker's realization is described by a sequence of three clauses. Semantically, each of these is coindexed with the "theme" argument of the previous; but syntactically, they are not embedded as complements, but rather linked anaphorically in a kind of topic chain. The first of these is marked for progressive aspect and talks about what the villagers were saying at topic time, which is still the time of the speaker's arrival. Because of the progressive, topic time is understood to fall into the time of the villagers saying this, so their talk is described as having started before the speaker's arrival. The final two clauses describe what the villagers were saying: they had heard on the radio that the hurricane was indeed going to hit their area. The clause referring to the villagers hearing the news on the radio carries the terminative AM marker, which functions much like a (tenseless) perfect: topic time is presented as falling into the post-state of the villagers hearing the news on the radio. In other words, they are said to have heard it before the speaker's arrival. The final clause refers to the approach of the hurricane. It is marked for progressive aspect, so the approach is presented as ongoing at topic time. It will be inferred that it was in fact already ongoing at the time this was announced on the radio, but this is not strictly entailed in (30).

The similarity to the English examples discussed in the introduction is intuitively obvious: some clauses introduce a new temporal perspective – in the present framework, the topic time – perhaps advancing the previous one, whereas others are interpreted with respect to this topic time. And the difference seems to depend, among other factors, on the aspectual properties of the clauses.

The existence of temporal anaphora in a tenseless language such as Yucatec is of course not surprising (although Bohnemeyer 1998b for Yucatec and Bittner (2008) for Kalaallisut are in fact the first descriptions of temporal anaphora in tenseless languages). Nevertheless, it is worth pointing out that in Partee 1973 and in the DRT literature adopting the concept, temporal anaphora is treated as part of the interpretation of tense markers in terms of a variable they introduce whose value is determined in context. The evidence from Yucatec and Kalaallisut makes it clear that utterances are interpreted with respect to topic times whether or not they are marked for tense, and that the determination of these topic times follows similar principles in tensed and tenseless languages.

In order to sketch an informal account of the interaction between temporal anaphora and aspect-mood marking, I would like to introduce the notion of the "Natural temporal reference point":

(31) **Natural temporal reference point (NTRP)**

A time interval *t* is an NTRP in a given discourse iff *t* is identified in that discourse as either (a) the coding time of some utterance or (b) a calendrical time interval or (c) an event time (the "run time" of an event described in the discourse).

This principle says that the times suitable as temporal reference points in discourse are the times identified on some calendar or clock-time scale and the times of events – including events described in the discourse and, in the case of deictic reference, the event of the production and/or comprehension of the utterance. What does *not* qualify a time interval as a suitable reference point or NTRP is the fact that some event is in progress at this time or that some state holds during it – for example a causal pre- or result state of some event, a state characterizing the realization of an event in some possible worlds, or a state characterizing the distance of the event from topic time. It follows that only perfective clauses, but not non-perfective clauses, introduce NTRPs.[20] Non-perfective clauses do not provide such reference points, but on the contrary require them for their interpretation. This is not to say that non-perfective clauses cannot be used to introduce topic time variables. In fact, this happens whenever a non-perfective clause is used to provide background information for a narrative sequence. An example is the stative topic clause in the second line of (29). Stative clauses are strongly preferred to be interpreted imperfectively. This means that the topic time introduced by the stative clause in (29) is not the entire time the man spent up in the tree, but rather some subinterval. If a perfective clause follows, that subinterval is understood to be a suitable time frame that contains the time of the event described by the perfective clause. The stative clause introduces the topic time variable for the perfective clause, but it is the inclusion of the

---

[20] In non-narrative discourses, clauses formed with the imperfective AM marker and irrealis topic clauses may be used to introduce new reference points. Examples of the latter option are (10) and (19) above. It seems that in these cases, event realization and thus the introduction of NTRPs is treated inside the scope of modal or habitual/generic operators.

event time of the perfective clause in this topic time that determines its value sufficiently to make it an NTRP. Suppose now instead of the perfective clause describing the perception of the demon's approach, the text would continue with the final progressive clause of (29), which describes the demon's approach as being in progress at topic time, as suggested in (32):

(32) Ti', bin, yàan te=ka'nal=o',
 there HS EXIST(B3SG) PREP:DET=high=D2
 'There he was, they say, up high (i.e., in a tree),'

 le=ba'l=o'; túun tàal
 DET=thing=D2 PROG:A3 come
 'the thing (i.e., the demon), it was coming'

The discourse in (32) might serve as background for a third clause using the perfective to place some event into the topic time of which we so far know that it falls into both the time of the man's being in the tree and the time of the demon's approach being in progress. But as a self-contained episode description, (32) not only makes a poor narrative ("nothing ever happens!"); it is also difficult to interpret. In (32), the imperfectively interpreted stative clause is combined with a progressive clause, which is semantically likewise imperfective. The progressive clause requires selection of a suitable topic time interval that falls into the time of the demon's approach. Even assuming the two topic times are identical, neither clause provides a suitable reference point for its resolution. Are the two states of affairs coextensive? Was the run time of one included in that of the other? Or did the two overlap partially? Principle (33) formulates the role of NTRPs in the selection of topic times:

(33) **Preferred topic time selection**

The topic times selected in a given discourse context are preferred to be identical to or include NTRPs identified in the same discourse context.

The possibility of inclusion of the NTRP in the topic time is mentioned in (33) because the topic times of perfective clauses are assumed in the present framework to contain the run times of the events described by the clauses (see Figure 1 in section 2.3). Principle (33) can be understood as a constraint on coherent discourses. Speakers craft their discourses so as to satisfy this constraint. The means at their disposal to manipulate topic time selection vary somewhat from language to language. In Yucatec, these are mainly viewpoint aspect and modal operators and adverbials; in English and other Indo-European languages, they also include tenses and temporal connective constructions.

Principles (31) and (33) together account for two key differences in the discourse behavior of perfective and non-perfective clauses. Non-perfective clauses trigger what may be called **binding implicatures** to the effect that their topic times are identical to some salient NTRP accessible in context. This can be coding time, giving rise to deictic interpretations, the time specified by some calendrical adverbial, or the event time of a perfective clause in surrounding discourse. Perfective clauses trigger no such implicatures because their topic times already include NTRPs – the run times of the events they describe. Perfective clauses can be used to introduce reference points "binding" the topic times of non-perfective clauses, and in the context of other perfective clauses, they may trigger the well-known **referential shift** interpretations. Consider (34):

(34) Káa=h-tàal-ech   way h-ts'o'k   ka'-p'éel ha'b=e',…
     CON=PRV-come-B2SG here PRV-end(B3SG) two-CL.IN year=TOP
     '(When) you came here two years ago,…'

   a. …káa=t-in=mèet-ah   le=nah=o'
      CON=PRV-A1SG=do:APP-CMP(B3SG) DET=house=D2
      '…I built the house'

   b. …táan in=mèet-ik   le=nah=o'
      PROG A1SG=do:APP-INC(B3SG) DET=house=D2
      '…I was building the house'

   c. …ts'o'k in=mèet-ik   le=nah=o'
      TERM   A1SG=do:APP-INC(B3SG) DET=house=D2
      '…I had built the house'

   d. …mukah in=mèet   le=nah=o'
      PROSP A1SG=do:APP(SUBJ)(B3SG) DET=house=D2
      '…I was going to build the house'

The clauses in (34a–d) follow the same perfective topic clause in the first line of (34). The perfective clause in (34a) is interpreted with respect to a new, "shifted" topic time not identical to the event time of the topic clause. Pragmatic inferences to be discussed below will locate the former just after the latter, giving rise to the interpretation that the speaker started building the house upon the addressee's arrival. In contrast, the continuations using progressive, terminative, and prospective aspect markers in (34b–d) are understood with respect to the event time of the topic clause as their topic time, giving rise to the interpretations that the construction was in progress (b), completed (c), or being planned (d) at the time of the addressee's arrival.

Binding implicatures are stereotype implicatures of the kind discussed by Atlas & Levinson 1981, generated by Grice's second Quantity maxim ("Do not make your contribution more informative than is required").[21] Referential shift, on the other hand, is the product of a combination of the failure to trigger binding implicatures and another stereotype implicature, this time to iconicity, i.e., to the effect that the order of clauses iconically reflects the order of events. With non-perfective clauses, this **iconicity implicature** is overridden by the binding implicature, since the latter is more specific – it yields "binding" of the topic time by a specific contextually accessible NTRP, whereas the iconicity implicature is satisfied by any time interval following the topic time of the preceding clause.

As generalized conversational implicatures, the binding and iconicity implicatures are defeasible default interpretations (Levinson 2000) triggered by the use of viewpoint-aspectual and modal operators in suitable contexts. The stative clause in (29) in fact illustrates blocking of the binding implicature: its topic time is interpreted to be shifted vis-à-vis that of the initial perfective clause because encyclopedic knowledge suggests that the man cannot have been sitting in a tree at the time he left for his *milpa*. A similar effect is illustrated in (35): the binding implicature is blocked because a balloon cannot be continued to be inflated once it has burst.

(35) Táan u=p'uru's-t-ik=e', káa=h-xíik-ih
 PROG A3=inflate-APP-INC(B3SG)=D3 CON=PRV-burst-CMP(B3SG)
 'She was inflating (the balloon), (when) it burst'

Defeasibility of the iconicity implicature can be illustrated by (12)–(13) in section 2.5, repeated here for convenience:

---

[21] If perfective clauses triggered binding implicatures, too, this would make narrative progression by referential shift impossible. Narratives would then require every perfective clause to be equipped with some device – an adverbial or some other expression – that explicitly signals "updating" of the topic time variable. As a matter of fact, this is precisely the function of the connective *káa* in the examples above, as discussed in section 2.5. The reason the perfective AM marker of Yucatec is accompanied by this connective in narrative discourse is that it is semantically vague between proper perfective (event time included in topic time) and perfect-like result-state interpretations (topic time included in the time of the result state).

(36) Pedro=e'   káa=t-u=ts'íib-t-ah   hun-p'éel
     Pedro=TOP CON=PRV-A3=write-APP-CMP(B3SG) one-CL.IN

   kàarta=e', káa=t-u=ts'u'uts'-ah   hun-p'éel chamal
   letter=TOP CON=PRV-A3=suck-CMP(B3SG) one-CL.IN cigarette

   'Pedro, (when) he wrote a letter, he smoked a cigarette'
   (preferred interpretation sequential)

(37) Káa=t-u=ts'íib-t-ah   hun-p'éel kàarta
     CON=PRV-A3=write-APP-CMP(B3SG) one-CL.IN letter

   Pedro=e', Juán=e', káa=t-u=ts'u'uts'-ah
   Pedro=TOP Juán=TOP CON=PRV-A3 suck-CMP(B3SG)

   hun-p'éel chamal
   one-CL.IN cigarette

   '(When) Pedro wrote a letter, Juán smoked a cigarette'
   (preferred interpretation for most consultants simultaneous)

The sequence of two perfective clauses in (36) triggers the familiar shift interpretation, while the one in (37) fails to do so because the two actions described have different agents. The discourse in (37) is certainly compatible with a sequential interpretation; it's just that out of context, most speakers consider the interpretation according to which the two actions occurred at the same time more salient. Bittner (2008), in her analysis of temporal anaphora in Kalaallisut within her "online update" framework, claims that temporal anaphora is in fact monotonic in "aspectually fully explicit" languages such as Kalaallisut. The examples presented above suggest that this analysis does not apply to Yucatec, even though Yucatec may well be considered "aspectually fully explicit", at least as far as the grammar of event descriptions, as opposed to state descriptions, is concerned.

The Gricean analysis of the inferences involved in aspect-based temporal anaphora resolution has the advantage over the standard DRT account that it offers an explanation for why these inferences and no others are triggered by aspectual operators and that it predicts the conditions under which these inferences are cancelled or blocked. The standard DRT approach accommodates the defeasibility of aspect-based temporal anaphora resolution as vagueness, based on the "event structure" model of Kamp 1979: the temporal reference of utterances is interpreted with respect to "instants", which in turn are constituted by sets of pairwise overlapping events. Thus, for example, while (34a) is said to introduce a new reference point at an instant that follows the reference point of the first clause, whereas (34b) is interpreted with respect to the reference point of the first clause, the two events may in both cases either overlap or follow one another. What this approach fails to explain is the default character of the inferences: sequential ordering in (34a) and simultaneity in (34b) will be inferred *unless* these interpretations are blocked or cancelled in context, as in (37) and (35). This default character follows from aspect-based temporal anaphora being rooted in generalized conversational implicatures, i.e., utterance-type meanings.

The account of temporal anaphora sketched above straightforwardly generalizes to the deictic uses of the aspect-mood markers. The binding implicatures triggered by non-perfective clauses are satisfied by calendrical adverbials or salient reference points available in context; where these are absent coding time takes over as the NTRP in accordance with (31) and (33). This gives rise to the deictic interpretations discussed in section 3: topic time is (or includes) coding time and is itself included in the run time of the event under description (with the progressive AM marker), a result state (with the terminative AM marker), a pre-state (with the prospective AM marker), some state that characterizes the realization of the described event in possible worlds (with the modal AM markers), or some state that characterizes the distance of the described event from topic time (with the "metrical" AM markers). Smith, Perkins, & Fernald 2007 note the same affinity of non-perfective clauses for deictic interpretations in Navajo, but propose a special "Deictic Principle" to account for it. In the present treatment, this affinity follows from the fact that coding time is a natural temporal reference point in combination with the semantics of non-perfective aspect-mood markers and the generalized conversational implicatures that govern temporal anaphora.

Clauses formed with the perfective AM marker are likewise used with topic time set to coding time, under result-state interpretations. Under perfective interpretations, they are excluded from deictic uses except for the marginal "blow-by-blow" online narrative context. This follows from the fact that the topic times of semantically perfective clauses include the run times of the events described by these clauses; these cannot be included in coding time except in the "blow-by-blow" scenario where coding time and topic time are continuously updated. In other words, all of deictic future and past time reference in Yucatec in fact involves present topic times (similarly Bittner 2005 for Kalaallisut).

5. Concluding remarks

Why do some languages have tense marking whereas others lack it? In my opinion, for essentially the same reason some languages (including English, Ewe, German, and Yucatec) mark their noun phrases for definiteness while the speakers of others (Estonian, Latin, Mandarin, Russian, Korean, and many more) seem to get by just fine without this device. Similarly, in some languages, noun phrases are marked for noun class or gender or trigger noun class or gender agreement (e.g., German, Kinyarwanda, Latin, and Russian) whereas gender and noun class play no role in the functional category system of other languages (e.g., in English, Estonian, Ewe, and Yucatec). In many languages, event descriptions are obligatorily marked for their viewpoint aspect (e.g., English, Ewe, Russian, Yucatec); but in some, they are not (e.g., Estonian and German). In languages such as Quechua and Turkish, clauses are obligatorily marked for the source of information the speaker purports to rely on, whereas in all other languages mentioned above, this is merely optionally indicated by lexical means. The reason for this kind of crosslinguistic variation in the functional category system seems to be that the expression of functional categories such as tense, viewpoint aspect, definiteness, gender, noun class, and evidentiality is not necessary for conveying the intended communicative content of linguistic utterances. The relevant conceptual distinctions are made whether or not they are expressed linguistically and speakers can rely on pragmatic means to communicate them where needed. Where functional categories *are* expressed, they serve to disambiguate and to facilitate reference resolution. One can thus surmise that there is a certain division of labor between pragmatics and the functional category system and a tradeoff between expressed and unexpressed categories. The expression of a rich system of aspectual and modal distinctions and simultaneous absence of tense marking in Yucatec exemplifies this tradeoff.

Acknowledgements

I am indebted to my Maya teachers and consultants, above all to Sebastián Baas May, Vicente Ek Catzin, Fulgencio Ek Ek, Amalia Ek Falcón, Máximo Ek Pat, Ernesto May Balam, Ramón May Cupul, Antonio May Ek, Saturnino May Ek, Norma May Pool, and Justina Pat May. I would like to thank the editors, the participants of SULA 4, and the members of the Buffalo Syntax/ Semantics Lab for helpful comments and suggestions.

References

Ayres, Glenn and Barbara Pfeiler
- 1997 *Los Verbos Mayas: La Conjugación en el Maya Yucateco Moderno* [Mayan Verbs: Conjugation in Modern Yucatec Maya]. Mérida, Yucatan: Ediciones de la Universidad Autónoma de Yucatán.

Abusch, Dorit
- 1985 *On verbs and times*. Doctoral dissertation, University of Massachusetts, Amherst.

Bertinetto, Pier Marco and Mario Squartini
- 1995 An attempt at defining the class of 'gradual completion verbs'. In *Temporal reference. Vol. 1: Semantic and syntactic perspectives*, P. M. Bertinetto, V. Bianchi, J. Higginbotham, and M. Squartini (eds.), 11–26. Torino: Rosenberg & Sellier.

Bittner, Maria
- 2005 Future discourse in a tenseless language. *Journal of Semantics* 22: 339–387.
- 2007 Online update: Temporal, modal, and de se anaphora in polysynthetic discourse. In *Direct Compositionality*, C. Barker and P. Jacobson (eds.), 363–404. Oxford: Oxford University Press.
- 2008 Aspectual universals of temporal anaphora. In *Theoretical and Crosslinguistic Approaches to the Semantics of Aspect*, S. Rothstein (ed.), 349–385. Amsterdam: John Benjamins.

Blair, Robert W. and Refugio Vermont Salas
- 1965–67 *Spoken (Yucatec) Maya*. Chicago: University of Chicago, Department of Anthropology.

Bohnemeyer, Jürgen
- 1998a Temporal reference from a Radical Pragmatics perspective: Why Yukatek does not need to express 'after' and 'before'. *Cognitive Linguistics* 9(3): 239–282.
- 1998b Time relations in discourse: Evidence from Yukatek Maya. Doctoral dissertation, Tilburg University.
- 1998c Die Stellung sententialer Topics im Yukatekischen (The position of sentential topics in Yucatec). In *Deskriptive Grammatik und allgemeiner Sprachvergleich* [Descriptive grammar and general language comparison], D. Zaefferer (ed.), 55–85. Tübingen: Niemeyer.
- 2000a Event order in language and cognition. In *Linguistics in the Netherlands* 17, H. de Hoop and T. van der Wouden (eds.), 1–16. Amsterdam: John Benjamins.
- 2000b Where do pragmatic meanings come from? The source of temporal inferences in discourse coherence. In *Samenhang in diversiteit. Opstellen voor Leo Noordman* [Unity in diversity. Papers presented to Leo Noordman], W. Spooren, T. Sanders, and C. van Wijk (eds.), 137–153. Tilburg: Tilburg University.

2001   Argument and Event Structure in Yukatek Verb Classes. In *The Proceedings of SULA: The Semantics of Under-Represented Languages in the Americas*, J.-Y. Kim and A. Weerle (eds.), 8–19. Amherst, MA: GLSA (University of Massachusetts Occasional Papers in Linguistics 25).
2002   *The grammar of time reference in Yukatek Maya*. Munich: Lincom Europa.
2003   Invisible time lines in the fabric of events: Temporal coherence in Yucatec narratives. *Journal of Linguistic Anthropology* 13(2): 139–162.
2004   Split intransitivity, linking, and lexical representation: the case of Yukatek Maya. *Linguistics* 42(1): 67–107.
2007   Morpholexical Transparency and the argument structure of verbs of cutting and breaking. *Cognitive Linguistics* 18(2): 153–177.
in press   Linking without grammatical relations in Yucatec: Alignment, extraction, and control. In *Issues in functional-typological linguistics and language theory*, Y. Nishina, Y. M. Shin, S. Skopeteas, E. Verhoeven, and J. Helmbrecht (eds.). Berlin/New York: Mouton de Gruyter.

Bohnemeyer, Jürgen and Penelope Brown
2007   Standing divided: Dispositionals and locative predications in two Mayan languages. *Linguistics* 45(5–6): 1105–1151.

Bohnemeyer, Jürgen and Mary D. Swift
2004   Event realization and default aspect. *Linguistics and Philosophy* 27(3): 263–296.
2006   Force dynamics and the progressive. Annual Meeting of the Linguistic Society of America; Albuquerque, NM.

Bricker, Victoria R.
1979   Wh-questions, relativization, and clefting in Yucatec Maya. In *Papers in Mayan linguistics 3*, L. Martin (ed.), 107–136. Columbia, MO: Lucas Brothers.

Comrie, Bernard
1976   *Aspect*. Cambridge: Cambridge University Press.
1985   *Tense*. Cambridge: Cambridge University Press.

Dahl, Östen
1984   Temporal distance: Remoteness distinctions in tense-aspect systems. In *Explanations for language universals*, B. Butterworth, B. Comrie, and Ö. Dahl (eds.), 105–122. Berlin/New York: Mouton de Gruyter.
1985   *Tense and aspect systems*. Oxford: Blackwell.

Dowty, David R.
1979   *Word meaning and Montague Grammar*. Dordrecht: Reidel.

Freed, Alice F.
1979   *The semantics of English aspectual complementation*. Dordrecht: Reidel.

Gordon, Raymond G., Jr. (ed.)
2005   *Ethnologue: Languages of the World.* 15th edition. Dallas, TX: SIL International.

Gutiérrez Bravo, Rodrigo
2006   Foco de agente y cláusulas relativas en maya yucateco [Agent focus and relative clauses in Yucatec Maya]. Manuscript, CIESAS-DF.

Hay, Jennifer, Christopher Kennedy, and Beth Levin
1999   Scalar structure underlies telicity in 'degree achievements'. In *SALT IX*, T. Mathews & D. Strolovitch (eds.), 127–144. Ithaca: CLC Publications.

Hinrichs, Erhard
1981   *Temporale Anaphora im Englischen.* M.A. dissertation, University of Tübingen.
1986   Temporal anaphora in discourses of English. *Linguistics and Philosophy* 9: 63–82.

Horn, Laurence R.
1972   *On the semantic properties of logical operators in English.* Mimeo. Bloomington, IN: Indiana University Linguistics Club.

Jackendoff, Ray S.
1983   *Semantics and cognition.* Cambridge, MA: MIT Press.

Kamp, Hans
1979   Events, instants, and temporal reference. In *Semantics from different points of view*, R. Bäuerle, U. Egli, and A. von Stechow (eds.), 376–418. Berlin: Springer.

Kamp, Hans and Uwe Reyle
1993   *From discourse to logic: Introduction to modeltheoretic semantics of natural language, formal logic and Discourse Representation Theory.* Dordrecht: Kluwer.

Kamp, Hans and Christian Rohrer
1983   Tense in Texts. In *Meaning, use, and interpretation of language*, R. Bäuerle, C. Schwartze, and A. von Stechow (eds.), 250–269. Berlin/ New York: Mouton de Gruyter.

Klein, Wolfgang
1994   *Time in language.* London: Routledge.
2006   On finiteness. In *Semantics in acquisition*, V. van Geenhoven (ed.), 245–272. Dordrecht: Springer.

Kratzer, Angelika
1977   What 'must' and 'can' must and can mean. *Linguistics and Philosophy* 1: 337–355.
1981   The notional category of modality. In *Words, worlds, and contexts*, H. J. Eikmeyer and H. Rieser (eds.), 38–74. Berlin/New York: Mouton de Gruyter.

1991 Modality. In *Semantics: An international handbook of contemporary research*, A. von Stechow and D. Wunderlich (Eds.), 639–650. Berlin/ New York: Mouton de Gruyter.

Krifka, Manfred
1992 Thematic relations as links between nominal reference and temporal constitution. In *Lexical matters*, I. A. Sag and A. Szabolcsi (eds.), 29–54. Menlo Park, CA: CSLI/SRI International.

Landman, Fred
1992 The progressive. *Natural Language Semantics* 1: 1–32.

Levin, Beth
1993 *English verb classes and alternations*. Chicago, IL: University of Chicago Press.

Levin, Beth and Malka Rappaport Hovav
1995 *Unaccusativity*. Cambridge, MA: MIT Press.

Levinson, Stephen C.
2000 *Presumptive meanings*. Cambridge, MA: MIT Press.

Li, Charles N. and Sandra A. Thompson
1981 *Mandarin Chinese: A functional reference grammar*. Berkeley, CA: University of California Press.

Lucy, John A.
1994 The role of semantic value in lexical comparison. *Linguistics* 32: 623–656.

Partee, Barbara
1973 Some structural analogies between tenses and pronouns in English. *The Journal of Philosophy* 70: 601–609.
1984 Nominal and temporal anaphora. *Linguistics and Philosophy* 7: 243–86.

Portner, Paul
1998 The progressive in modal semantics. *Language* 74: 760–787.

Reichenbach, Hans
1947 *Elements of Symbolic Logic*. New York, NY: Free Press.

Shaer, Benjamin
2003 Toward the tenseless analysis of a tenseless language. J. Anderssen, P. Menéndez Benito, and A. Werle (eds.), 139–56. *Proceedings of SULA 2*. Amherst, MA: GLSA.

Smith, Carlota S.
1978 Jespersen's 'move and change' class and causative verbs in English. In *Linguistic and Literary Studies in Honor of Archibald A. Hill, Vol. 2: Descriptive Linguistics*, M. A. Jazayery, E. C. Palome, and W. Winter (eds.), 101–109. The Hague: Mouton.

Smith, Carlota S., Ellavina T. Perkins, and Theodore B. Fernald
2007 Time in Navajo: Direct and indirect interpretation. *International Journal of American Linguistics* 73 (1): 40–71.

Tonhauser, Judith
    2003   F-constructions in Yucatec Maya. In *Proceedings of SULA 2*, J. Anderssen, P. Menéndez Benito, and A. Werle (eds.), 203–223. Amherst, MA: GLSA.

Van Valin, Robert D. Jr. and Randy J. LaPolla
    1997   *Syntax*. Cambridge: Cambridge University Press.

Vapnarsky, Valentina
    1995   Las voces de las profecías: Expresiones y visiones del futuro en maya yucateco (The voices of the prophesies: Expressions and visions of the future in Yucatec Maya). *Trace* 28: 88–105.
    1999   *Expressions et conceptions de la temporalité chez les mayas yucateques (Mexique)* (Expressions and concepts of temporality among the Yucatec Mayans [of Mexico]). Ph.D. dissertation, Department of Ethnology and Comparative Sociology, Université de Paris X Nanterre.

Tenses in compositional semantics

Arnim von Stechow

1. Aim of this article

Most scholars assume that the Present in matrix clauses denotes the speech time, but there is little agreement on the meaning of the Past. I am aware of at least three approaches:

(i) Past is an existential quantifier that instantiates the embedded VP to some time before the speech time; this view is attributed to tense logic, e.g. Prior (1967).[1]

(ii) Past is a referring term denoting some contextually salient time before the speech time; this semantics is attributed to Partee (1973).[2]

(iii) Past is a predicate that applies to a time t and says that t is before the speech time; this view originates perhaps with Dowty (1979).[3]

I think that each of these approaches can be made to work with auxiliary assumptions. Complications will arise, however, as soon as we study the semantics of Tense in embedded constructions. Here tenses will either contain a bound variable or they are simply bound variables and none of the above approaches can be applied.

To see what is necessary for a successful semantic treatment of Tense therefore requires the study of Tense in subordinate constructions such as relative clauses, verbs of attitude and temporal adjunct clauses. Furthermore we have to make sure that Tense interacts in a correct way with temporal adverbs, with negation, and quantification.

In this paper I will follow the approach in Heim (1997), which is in the spirit of tense logic. The approach will be combined with a theory that has subordinate Tense as a bound pronoun. Combined with a theory of feature

---

[1] The indefinite approach is used in Ogihara (1989) and Kusumoto (1999), among others.
[2] Followers of this approach are Abusch (1994) and Kratzer (1998), among many others.
[3] Systems along these lines are found in von Stechow (1995) and Musan (2002) among others.

transmission under binding (cf. e.g. Heim 2005), the approach, I believe, contains the essential ingredients for an adequate theory of tense in English and other languages. I will also refer to a number of other proposals made in the literature; but it should be clear that in such a short paper, it is impossible do justice to the entire literature on tense in compositional semantics; there may be other work of equal importance.

The structure of the article is as follows. Section 2 introduces the ontology of time together with some terminology. Natural language requires a distinguished time s*, the speech time ("origo", see Klein 2008). Times will be intervals composed of moments. Section 3 is about the relation of predicates of natural language to time. I will assume that each basis predicate has exactly one time argument or an argument of something that uniquely determines a time. I will illustrate this claim by a discussion of the Vendler Aktionsarten. Section 4 introduces an extensional semantic framework; the intensional version is introduced in section 11.2, where attitudes are treated. Section 5 introduces the semantics for Present and Past and those of the temporal auxiliaries. Section 6 deals with the syntax/semantics interface: a semantic tense transmits its feature to a finite verb under binding. Section 7 treats the interaction of Tense with temporal adverbs such as *yesterday*, *at six o'clock* and asymmetric temporal quantifiers like *on every Sunday*. Section 8 discusses the referential theory of tense and the Partee Problem; it defends an indefinite semantics for the Past. Section 9 introduces contextual restriction for tense. Section 10 contains remarks on the scope interaction between Tense and quantifiers and negation. Section 11 is devoted to tense in subordinate constructions.

2. The ontology of time

For doing semantics it not necessary to say precisely what times are.[4] It is enough to make some structural assumptions that clarify how language deals with time. The great majority of semanticists cling to the idea of a time line that consists of ordered time intervals. The line is composed of moments that are ordered by the "before" relation <, which is a linear ordering, i.e. for two moments m_1, m_2 we have: $m_1 < m_2$ or $m_2 < m_1$ or $m_1 =$

---

[4] Personally, I think that times are equivalence classes of stretches of possible worlds that occur at the same time. (Kamp and Reyle 1993) hold the view that times are equivalence classes of events that occur at the same time. This is much the same.

m_2. The converse relation "after" is denoted by $>$, i.e., $t > t'$ iff $t' < t$. Most semanticists assume that time is continuous, i.e., the moments are similar to real numbers (see, e.g. Klein 2008). Dowty (1979) assumes discrete time because the definition of his BECOME-operator, which is used for the analysis of achievements/accomplishments, requires that for each time there is exactly one next time. Section 3.2 shows a way to define achievements in a continuous time, so time can be continuous.

The $<$-relation is extended in a natural way to intervals: The interval t is before the interval t' if each moment in t is before any moment in t'. Time intervals t are assumed to be "convex" in the following sense: if m_1 is the first moment in t and t and m_2 is the last moment in t, then any moment between m_1 and m_2 is in t as well. This terminology assumes that time intervals are closed, i.t. that each t has the form $[m_1, m_2] = \{m \mid m_1 \leq m \leq m_2\}$, where m_1 is the left bound of t – lb(t) – and m_2 is the right bound of t – rb(t).

Intervals may overlap, which is represented by O, or they may stand in the subinterval relation ($\subseteq$) or proper subinterval relation $\subset$. And, of course we may form the union $t \cup t'$ of the two times t and t', where t and t' should be contiguous or overlapping. Similarly, we can intersect two times i.e., form the $t \cap t'$ from the times t and t'. The relations O and $\cup$ may be defined on the basis of $\subseteq$.

Time spans have length/duration, which is measured by means of an appropriate unit: second, hour, day. Lengths satisfy the usual conditions for distances. In this article duration won't play a role.

The notion of tense in natural language is a deictic category, i.e., a tense is always related to the deictic centre, the speech time s* (called *origo* in Klein 2008).

Taken together, we are assuming the following *time structure*:

(1) $<M, T, <, \subseteq, QU, s^*>$

where M are the time points, T (the time spans) is the set of closed intervals formed from time points; $<$ is the before relation defined both for points and for intervals, $\subseteq$ is the subinterval relation, QU is a function assigning to each time interval a length, s* is the speech time.[5]

---

[5] The notation s* for the speech time is used by many authors, e.g. (Ogihara 1996), QU for time measure is used in (Kamp and Reyle 1993).

3. Vendler Aktionsarten and the time argument

In this section I want to clarify the following questions:

1. Do we need points of time/moments? (Yes.)
2. Is time discrete or a continuum? (Possibly a continuum.)
3. How many time arguments has a predicate? (Depends. One or two.)

I try to give answers by recapitulating what has been said in the literature about the temporal properties of Vendler Aktionsarten.

Every semanticist working on tense assumes different temporal properties for the so-called *Vendler Aktionsarten* for verbs and adjectives: States, Achievements, Accomplishments, and Activities; see Vendler (1957). There is agreement that these properties are not properties of verbs (or adjectives) in isolation but of VPs. Which Aktionsart is expressed by a VP is compositionally determined from the meanings of its parts. Nevertheless, the temporal structure of the lexical entry should be appropriate for the composition. For the following definitions we assume temporal properties P of type it.[6]

3.1. States

States are typically expressed by adjectives. The semantics of states poses a serious problem for the now popular idea that language only speaks about time intervals. We will say that states are predicates of moments. Look first at the following classic definition from Tailor (1977).

(2) **States.** The temporal property P is a *state* (or stative) if for every time t:
$P(t) \leftrightarrow (\forall t' \subseteq t)\, P(t')$

In other words, states have the *subinterval property*. The tenseless VP **the shop be open** (called lexical content by Klein 2008) is a state: if the shop is open at the interval t = [9 a.m., 7 p.m.], then the shop is open at any subinterval thereof. The condition raises the question of what the meaning of **open** should be. There are two options: (i) the subinterval property could be implemented in the meaning; (ii) the adjective could be defined for time moments and a covert logical operation could extend the predicate to intervals.

---

[6] Types are introduced systematically in section 4.

(3) **open**
   a. [[**open**]] = $\lambda t \in T. \lambda x.(\forall t' \subseteq t)$ x is open at t'. (interval semantics)
   b. [[**open**]] = $\lambda m \in M. \lambda x.$ x is open at m. (point semantics)

Consider the LF of the VP under the first alternative.

(4) [[**the shop open**]] = $\lambda t.(\forall t' \subseteq t)$ the shop is open at t'.

This meaning raises the question of how it combines with negation. Consider the VP **the shop be not open** and abbreviate this property as ϕ. Adopting the interval semantics (3a), the negated VP ¬ϕ is true of an interval t if ϕ is false of at least one subinterval of t. This is a very weak condition. Consider the following situation with abutting intervals t_1 and t_2:

```
    t₁        t₂
|---------|---------|
    ϕ        ¬ϕ
         |_____| t₃
           ¬ϕ
```

The subinterval property entails that ϕ is true at $rb(t_1) = lb(t_2)$. The semantics predicts that ¬ϕ is true at the interval t_3, though at the first half of t_3 ϕ is true. ¬ϕ would even be true of t_3 if t_3 were a final subinterval of t_1, i.e., only the last point of t_3 would make ϕ false.

Now, consider the VP **the shop is not open from 9 A.M. to 8 P.M.**. The prevailing reading is that the shop is not open, i.e. closed, throughout the interval [9 a.m., 8 p.m]. We cannot express this by means of the entry (3a). Suppose therefore that we have the covert adverb **THR**(oughout):

(5) [[**THR**$_{i(mt,t)}$]] = $\lambda t.\lambda P_{mt}.(\forall m \in t) P(m)$

The VP **the shop be not open** can have two logical forms:

(6) $\lambda t.$ **not**$_{tt}$ **THR**(t) **the shop open** = $\lambda t.\neg(\forall m \in t)$ the shop is open at m

This is the reading we get with the interval semantics.

(7) $\lambda t.$ **THR**(t) **not**$_{tt}$ **the shop open**
   $= \lambda t.(\forall m \in t)$ the shop is not open at m

This is the intuitively prevailing reading. Since I don't see how the second reading can be obtained from the interval semantics, I will assume that the mt ("moment to truth-value") semantics is the correct version.

States are typically expressed by (stage-level) adjectives: **sick, drunk, happy, asleep**,... Locative prepositions express states as well: **the book be on the table** means λt.the book is on the table throughout t. The so called object-level stative verbs, e.g. **know, like** express states as well (cf. Dowty 1979: 180): **Wolfgang like Eva** means λt.Wolfgang likes Eva at throughout t. Similarly locative verbs like **lie, stand, sit** may express states as well: **the Nile lie in Africa** means λt.the place of the Nile is in Africa throughout t. Finally, states are expressed by stage-level nouns: **Sue be a student** means λt.Sue is a student throughout t.

The conclusion is: the semantic analysis of states requires moments. States are what Klein (1994) calls 1-state predicates; each instance of such a predicate is a pair consisting of one individual and *one* moment. States are projected to time intervals via an additional operation **THR**. The latter point is not standard. In what follows I will ignore this insight and deal with states as if they where sets of intervals *simpliciter*, but a proper analysis requires the said decomposition.

3.2. Achievements

Achievements are predicates that describe sudden changes: **realize, discover, spot, find**. Intuitively, an achievement is true of a point of change between a ¬φ-state to a φ-state.

(8)   Achievements
      $----|++++$
      where $---- = \neg \phi$, $++++ = \phi$, $| =$ the point of change

The non-trivial question is how such points of changes can be described by lexical contents.

Consider the VP **Franzis find her wallet**. Intuitively this is true at a moment m if Franzis sees her wallet at m and she doesn't see it immediately before m. In a theory that works with discrete time we have no problem analysing this. Let us denote the predecessor of a moment m pred(m). The VP mentioned therefore means λm.Franzis sees her wallet at m & she doesn't see it at pred(m). The corresponding lexical entry would be this:

(9) $[[\textbf{find}_{m(e(et))}]] = \lambda m \in M.\lambda x.\lambda y.y$ doesn't see x at pred(m) & y sees x at m.

The first conjunct of the truth-condition should better be a presupposition, a refinement we neglect. So the semantics of achievements seem to require discrete time.

Dowty (1979) gives an interval semantics for achievements, which requires a discrete time structure as well. His formalisation uses the BECOME-operator, but to make the point, we can formulate the lexical entry of **find** directly:

(10) Dowty's discrete interval semantics for achievements

[[**find** $_{i(e(et))}$]] = $\lambda t \in T.\lambda x.\lambda y.y$ doesn't see x at lb(t) & y sees x at rb(t) & $\neg(\exists t' \subset t)$ y doesn't see x at lb(t') & y sees x at rb(t')

The reader should convince himself that any interval t that satisfies these conditions must consist of two moments.[7]

But we are not forced to the conclusion that the semantics for achievements requires discrete time. Here is a semantics for achievements that works for a continuous time structure, as far as I can see:

(11) Achievements for continuous time

[[**find** $_{m(e(et))}$]] = $\lambda m \in M.\lambda x.\lambda y.(\exists n_1 < m)[y$ doesn't see x at n_1 & $(\forall n_2)[n_1 < n_2 < m \rightarrow y$ doesn't see x at $n_2]]$ & y sees x at m

This means that the $\neg\phi$-interval approaches the ϕ-point from the left without ever reaching it. So the question of the predecessor or successor doesn't arise. The conclusion is that the semantics for achievements requires the existence of time points, but time may nevertheless be continuous. Let me say again that the first conjunct in the definition is better considered as a presupposition. As for states, I will simplify in the following by treating achievements as if their time argument were a short interval.

---

[7] Suppose t did consist of two short abutting subintervals t_1 and t_2, each consisting of more than one point. t_1 is a $\neg\phi$-interval, t_2 is a ϕ-interval. But then we could find two shorter subintervals t, t' fulfilling the conditions and so on down to two neighbouring moments. So t has the form $[m_1,m_2]$ with m_1 immediately before m_2. But in a continuum such an interval cannot exist because real number don't have successors.

3.3. Accomplishments

The property of Accomplishments accepted by everyone is:

(12) "Quantization"
P_{it} is an *Achievement/Accomplishment* if for any time t: If P(t) and t' is a proper subinterval of t, then $\neg P(t')$-

The term "quantization" comes from Krifka (1989), but the condition originates with Vendler. If achievements are only defined for moments, the property of quantization follows trivially. An *accomplishment* P is quantized and only defined for genuine intervals, i.e., intervals that contain more than one point. As an example consider the VP **Max polish his car**. This denotes the set of intervals t such that Max starts polishing his car at the beginning of t, he works on his car throughout t and the car is clean at the end of t. So the meaning of **polish** is something like this:

(13) $[[\text{ \textbf{polish}}_{i(e(et))}]] = \lambda t.\lambda x.\lambda y.y$ works on at t such that x is clean at rb(t)
  $= \lambda t.\lambda x.\lambda y.y$ polishes x at t

In this particular case the property of being an accomplishment is determined by the lexical semantics of the verb As has been mentioned earlier, this is not always so. The VP **John walk** is what is called an *activity* (see next section). But the VP **John walk from the Post Office to the station** is an accomplishment. The example is from Dowty (1979), who presents many other examples of this sort.

3.4. Activities

An activity is very similar to a state. An activity has the subinterval property down to some small intervals that are necessary to make the predicate true. The classical examples are **John walk** or **John waltz**: the activity of walking requires that at least one step is taken, and that of waltzing requires at least three steps. So these predicates cannot be true of moments. Activities are cumulative (or summative):

(14) Cumulativity of activities
If P_{it} is activity and P holds of two abutting intervals t_1 and t_2, then P holds of $t_1 \cup t_2$.

Since activities are almost states, they should be treated in a similar way, i.e., they should be defined for minimal intervals, which are extended by an appropriate version of the covert operator THR to larger interval.

(15) $[[\textbf{walk}_{i(et)}]] = \lambda t.\lambda x.x$ walks at t

Each t in this set contains only one minimal walk of the subject. But we will be sloppy by assuming that these walking intervals can be very long.

3.5. How many time arguments?

An inspection of the lexical entries for the different kinds of predicates given in the last section reveals that I have assumed one time argument for each predicate. Now, achievements and accomplishments are what Klein (1994) calls "2-state verbs". He holds the view that these verbs must have two time arguments; recently he claims that two may not even be sufficient; cf. Klein (2000, 2002). It is true that achievements and accomplishments are concerned with two times (the time before the change and the time after the change for achievements, the action time and the target time for accomplishments). But in each case the other time can be recovered from the unique time the predicate speaks about. A least, this is the usual assumption present in most work about tense semantics. However, one time argument is not sufficient for every purpose. For instance, the stative (or adjectival) passive seems to project the "target state" on the time axis. The VP **the car be polished** can mean "the car is clean as a result of a polishing". Clearly this is true of times that are immediately after a polishing; the car still has to be clean at those times. Suppose then that the lexical entry for **polish** is the following:

(16) **polish** as a 2-state verb: type t(m(e(et)))
 $[[\textbf{polish}]] = \lambda m.\lambda t.\lambda x.\lambda y.y$ polishes x at t and m = rb(t)

The "stativizer" that forms the adjectival passive would then be an operation that "absorbs" the event time and the agent and project the target time and the patient. Apart from details, this account goes back to Kratzer (2000):

(17) The stativizer ST: type (m(i(e(et))))(m(et))
 $[[\textbf{ST}]] = \lambda R_{m(i(e(et)))}.\lambda m.\lambda x.(\exists t)(\exists y)R(m)(t)(x)(y)$

Thus the adjectival passive VP has the LF **λm.the car [ST polished](m)** and means $\lambda m.(\exists t)(\exists y)[y$ polishes x at t and m = rb(t)]. When the verb is in the active, we need a different operation that "absorbs" the target time and projects the event time:

(18) The eventizer EV, type (m(i(e(et))))(i(e(et)))

[[EV]] = $\lambda R_{m(i(e(et)))}.\lambda t.\lambda x.\lambda y.(\exists m) R(m)(t)(x)(y)$

So the VP **Max polish his car** has the LF **λt.Max [EV polish](t) his car** and means $\lambda t.(\exists m)[m = rb(t)$ & John polishes his car at t]. This is exactly what we had under our 1-state treatment of **polish**.

The outcome of this discussion is that it is correct that verbs ultimately may have more than one time argument, but when it comes to the VP, where the subject and the object have been plugged in, only one argument remains. The other argument is bound by one of the two operations. In order to not complicate the picture, we will assume that verbs have only one time argument.

The question that still has to be answered is this: Which argument of the verb (or other predicates) is the time argument? Is it the first or is it the last? I know of no convincing evidence to decide this question. In this article, I will assume that the time argument is the first argument of the predicate.

The summary of this discussion is that states are sets of intervals, but stative verbs are sets of moments. We come to the states via the operation **THR**. These have the subinterval property. Achievements are sets of moments, which we identify with very short intervals. Accomplishments are sets of time intervals. Activities are somehow in between accomplishments and states. They have the subinterval property, but only down to some granularity.

4. Logical form

In this section I introduce an extensional typed language L that will serve for the representation of logical forms (LFs). The intensional extension is introduced in section 11.2.

The *types* of L are: *e* (entities), *i* (time intervals), *m* (moments), *s* (worlds) and *t* (truth-values). The functional types are generated by the following rule: if *a* and *b* are types, then *(ab)* is a type. Outermost brackets are usually omitted.

The syntax of L is based on a lexicon of expressions belonging to some type in the style of the examples already given. Furthermore, we have infinitely many variables for any type (often written as numbers with type indices or as traces with numbers). Finally we have the following syntactic rules:

(19) Syntax of L
   1. If α is a lexical entry or a variable of type a, them α is an expression of type a. ("Lexicon")
   2. If α is an expression of type (ab) and β is an expression of type a, then [α β] is an expression of type b. ("functor-argument")
   3. If α and β are expressions of type (at), then [$\alpha\beta$] is an expression of type (at). ("predicate modification")
   4. If α is an expression of type a and x is a variable of type b, then [$\lambda x \alpha$] is an expression of type (ba). ("abstraction")

The interpretation of the language is based on a model $\mathfrak{M} = (M, T, E, \{0,1\}, s^*, F)$, where M is the set of moments, T is the set of time intervals based on M, E is the set of entities, $\{0,1\}$ is the set of truth-values, and s* is the speech time. (A more inclusive model should include the set W of possible worlds.) F is a function that interprets the lexicon with appropriate meanings. To do this, we need a system of semantic domains D_a for each type a. We define: $D_m = M$, $D_i = T$, $D_e = E$, $D_t = \{0,1\}$, and $D_{(ab)}$ = the (possibly partial) functions from D_a into D_b. A condition for the interpretation of the lexicon is that $F(\alpha) \in D_a$ if α is a lexicon entry of type a.

We now define the function $[[\,.\,]]^{\mathfrak{m},g}$ that interprets each expression of L. The function depends on the model $\mathfrak{M}$ and a variable assignment g.

(20) Interpretation of L
   1. a. If α is a lexical entry of type a, then $[[\alpha]]^{\mathfrak{m},g} = F(\alpha)$. ("Lexicon")
      b. If x is a variable of type a, then $[[x]]^{\mathfrak{m},g} = g(x)$.   ("Variable")
   2. If α is of type ab and β is of type a, then
      $[[[\alpha\beta]]]^{\mathfrak{m},g} = [[\alpha]]^{\mathfrak{m},g}([[\beta]]^{\mathfrak{m},g})$.   ("Functional Application", FA)
   3. If α and β are of type at, then $[[[\alpha\beta]]]^{\mathfrak{m},g} = \lambda x_a.[[\alpha]]^{\mathfrak{m},g}(x)\ \&\ [[\beta]]^{\mathfrak{m},g}(x)$.   ("Predicate Modification", PM)
   4. If x is a variable of type a and a α is an expression of type b, then
      $[[[\lambda x \alpha]]]^{\mathfrak{m},g} = \lambda u \in D_a.[[\alpha]]^{\mathfrak{m},g[x/u]}$.   ("Abstraction", λ)

$g[x/u]$ is defined like g with the (possible) exception that $g[x/u](x) = u$. As a special case of PM we will assume that α and β are of type t. Then $[[[\alpha\beta]]]^{\omega,g} = 1$ iff $[[\alpha]]^{\omega,g} = 1 = [[\beta]]^{\omega,g}$.

5. Temporal structure of simple sentences

We start with the sentences:

(21) a. John called.
 b. Mary is happy.

There is a long tradition in Generative Grammar that the structure of finite sentences is $[_{TP} \text{ T VP}]$. In other words, a Tense is the head of the sentence. At S-structure, the subject moves to [Spec,TP] for case reasons, and the main verb moves to T. We ignore these movements for semantics. We will assume the following two Tenses for English:

(22) Tenses
 a. Present, type i: $F(\mathbf{N}) = s^*$.
 b. Past, type i(it,t): $F(\mathbf{P}) = \lambda t.\lambda P_{it}.(\exists t')[t' < t \ \& \ P(t')]$

Semantic tenses are covert, i.e. they are not pronounced. The semantic past **P** is a relative tense, it shifts the local evaluation time backwards. We assume that the argument of **P** is always **N** in matrix clauses. This warrants a deictic interpretation. (In subordinate clauses, P can have a time variable t as argument that is bound by a higher tense or locally bound by a λ-operator. Matrix Past is represented as [**P N**]. This is quantifier and requires an argument of type it. How is this possible? Here are the lexical entries for **John** and **called**:

(23) a. $F(\mathbf{John_e}) = \text{John}$
 b. $F(\mathbf{called_{i(et)}}) = \lambda t.\lambda x.x \text{ calls at t}$.

Note that **called** has a tenseless semantics. In other words the morphology of the verb is not directly reflected in its semantics. I will take up the morphology/semantics interface in the next section. The first argument of the verb is a time argument. At deep structure this is filled with a semantically empty pronoun PRO, which has no meaning and no type. Therefore it has to be moved for type reasons creating a λ-abstract of type it. At LF PRO is deleted. This is the PRO-theory of Heim and Kratzer (1998). Here is the derivation of the LF for sentence (21a); the movement of the subject to [Spec,TP] and the movement of the verb to T is ignored.

(21) DS: [$_{TP}$ [$_T$ P N] [$_{VP}$ John [called PRO]]]
PRO-movement (with subsequent PRO deletion)
LF: [$_{TP}$ [$_T$ **P N**] ~~PRO~~ λ$_1$ [$_{VP}$ **John** [**called t$_1$**]]]
= (∃t < s*) John calls at t

The LF is transparent, i.e. interpretable. The reader my compute for herself that it has the interpretation indicated. The method of calculation is entirely standard. We have to use Functional Application and the abstraction rule.

The analysis of sentence (21b) requires the introduction of the auxiliary **be**, which has an entirely trivial semantics: it simply passes the matrix tense to the adjective: adjectives have a time argument but they aren't finite forms.

(24) The temporal auxiliary **be**: type i(it,t)
F(**is**) = λt.λP$_{it}$.P(t)

Here is the derivation of the LF for the sentence:

(25) DS: [$_{TP}$ P N [$_{VP}$ [$_V$ is PRO] [$_{AP}$ Mary happy PRO]]]
PRO movement (with subsequent PRO deletion)
LF: [$_{TP}$ **P N** λ$_1$ [$_{VP}$ [$_V$ **is** t$_1$] λ$_2$ [$_{AP}$ **Mary happy** t$_2$]]]
= (∃t < s*) Mary is happy at t

Again, the reader may convince himself that the LF has the meaning indicated.

The Perfect and the Pluperfect require the auxiliary **have**, which has the same meaning as the semantic Past **P**.

(26) **have/had**: type i(it,t)
λt.λP$_{it}$.(∃t')[t' < t & P(t')]

Thus the analysis of *John had called* is the following:

(27) [$_{it,t}$ **P N**] [λ$_1$ [[had t$_1$][λ$_2$ [John [called t$_2$]]]]]
= (∃t$_1$)[t$_1$ < s* & (∃t$_2$ < t$_1$)[John calls at t$_2$]]

There is something idiosyncratic with the English Present Perfect, which requires a special meaning. The standard account for **have** under semantic

Present is that the auxiliary expresses an "extended now" XN; cf. Dowty (1979: ch. 7). This is an interval whose right bound is s*:

(28) XN-Perfect

$[[\text{ has }]] = \lambda t.\lambda P_{it}.(\exists t')[t \text{ is a final subinterval of } t' \ \& \ P(t')]$

The future auxiliary **will** is the mirror image of **have**:

(29) **will**: type i(it,t)

$\lambda t.\lambda P_{it}.(\exists t')[t' > t \ \& \ P(t')]$

(30) John will call.

N $[\lambda_1 \ [[\text{will } t_1][\ \lambda_2 \ [\text{John [call } t_2]]]]]$
$= (\exists t')[t' > s^* \ \& \ \text{John calls at } t']$

The English Progressive is a modal operator and requires the introduction of the world argument into the semantics. I give a hint what the analysis of *Max was polishing his car* will be. The auxiliary **was** expresses Dowty's (1979) PROG-operator. In other words, the auxiliary **be** is ambiguous between a tense transmitter and the Progressive. The LF will be this:

(31) $[_{TP} \ \textbf{P} \ \textbf{N}_1 \ [_{VP} \ \textbf{was}(@,t_1) \ \lambda_2 \lambda w \ [_{VP} \ \textbf{Max polishing}(w,t_2) \ \textbf{his car}]]]$
$= (\exists t < s^*)(\forall w)[@ \ R_t \ w \rightarrow (\exists t')[t \subseteq t' \ \& \ rb(t') > rb(t) \ \& \ \text{Max polishes his car in } w \text{ at } t']]$

@ is the actual world and $@ \ R_t \ w'$ means that w is accessible to @ at time t. R must be an appropriate accessibility relation, more or less what Dowty calls the "inertia" worlds. These have a common past with the real world but might diverge from the real world in the future but should be compatible with our expectations at time t. There is a large amount of literature on the Progressive. Notice that the progressive verb form **polishing** has exactly the same meaning as other forms of the verb. PROG is not expressed by this verb but by the auxiliary **was**.

Some remarks on the literature are in order. Semantic Past is decomposed into a relative tense (like the Perfect) and the deictic Present (or a time variable). As far as I know this decomposition was been proposed for the first time in Heim (1997). It is also used in Kusumoto (1999). Our semantic tense is indefinite, i.e. **P** is a restricted existential quantifier. In many places in the literature we find a relative Past that quantifies over an implicit time parameter on which the interpretation depends: $[[\ \textbf{P} \ \alpha \ ]]^t = (\exists t' < t)$

[[α]] [t]; see for instance Montague (1973). Most authors (e.g. Ogihara 1996; Kusumoto 1999) attribute this semantics to Prior (1967). I have not been able to verify this; Prior's work is entirely axiomatic and mostly concerned with the reconstruction of Diodorian modality. But the interpretation seems to be in the spirit of Prior. **N** is the "now" of Kamp (1971). The great advantage of a decomposed Past is that it can correctly deal with back shifted readings in subordinate constructions. Note that other languages have a different interpretation for the Present. In Japanese and German the Present can express a non-Past, i.e. $\lambda t.\lambda P_{it}.(\exists t')[\neg t' < t \ \& \ P(t')]$ or some related, even more complicated versions using overlap with the speech time; see Abusch (1994), Klein (1994), Musan (2002) among others for different variants. We have seen that **have** may have different meanings, either it is synonymous with **P** or it expresses an XN. Musan (2002) defends the view that the German Perfect gives us a time that is either before a reference time t or gives us an XN in the limiting case, i.e. Perfect = $\lambda t.\lambda P.(\exists t')[t' < t$ or t' is a final subinterval of t & $P(t')]$. The auxiliary system proposed in this section is compositional. This is a controversial matter.[8] For a survey and discussion of the different proposals that have been made, see Kratzer (1998).

6. Syntax and morphology

I am assuming a feature theory in the style of Zeijlstra (2004). There are two sorts of features, interpretable ones [iF] and uninterpretable ones [uF]. The interpretable features originate with certain logical operators, e.g. Negation or Tense. They reflect the meanings of these operators. The uninterpretable features are carried by other expressions that do not have the said meaning but may morphologically expose the meaning. An interpretable feature may check an uninterpretable one under agreement. If one i-feature checks more than one u-feature, we have "Multiple-Agree".

In English, finite verb forms have uninterpretable temporal features. Present forms of a verb a have the feature [uN] "uninterpretable Present/Now". The semantic Present **N** has the feature [iN] "interpretable Present/Now". Past forms of a verb have the feature [uP] "uninterpretable Past". The semantic Past tense **P** has the feature [iP].

---

[8] For instance, clausal participles require covert operators (*die vor Jahrunderten erbaute Burg* 'the before centuries built castle', *die seit Jahrhunderten zerstörte Burg* 'the since centuries destroyed castle'). Some researchers think that the semantic Past is contained in the participle and auxiliaries are semantically void.

144 *Arnim von Stechow*

(32) Some verb forms with spell out:
Present: **call/calls** [uN]
Past: **called** [uP]
Past Participle: **called** (no temporal feature)
Infinitive: **call** (no temporal feature)
Meaning of all these: λt.λx.x calls at time t

We will assume the following conventions for the transmission of temporal features under binding.

(33) Feature transmission under semantic binding.

A semantic tense **P** or **N** transmits a feature [uP]/[uN] to the time variable it binds. If the variable is an argument of a tensed verb form, the feature has to agree with the tense feature of the verb.

The idea that features may be transmitted under semantic binding is from to Irene Heim; see e.g. Heim (1994a, 2005).

We assume the conventions for semantic binding outlined in Heim and Kratzer (1998). In particular, a phrase or operator α may bind a variable via a λ-operator. To generate the necessary λ-operators, we assume that caseless argument positions are filled with the semantically empty pronoun PRO, which must be moved ("QR-ed") thus creating a λ-abstract. Being semantically empty, PRO is deleted after movement. Case positions may be filled with the semantically empty operator WH, the relative pronoun. It is moved and creates a λ-abstract as well. Here is the analysis of a simple sentence.

(34) Mary called.

Deep structure (=DS)

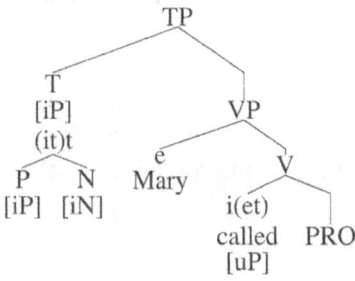

(The subject **Mary** is moved to [Spec,TP] for case reasons at SS. This will be ignored.) The tree assumes the following convention for feature percolation:

(35) Percolation of tense features
   a. Features percolate along the head line.
   b. The feature of a temporal variable either agrees with the inherent feature of the head or it is transmitted to the head (and percolates to the phrase).

Since the semantic Past is the head of the semantic tense [P N], the feature [iP] percolates to the phrase [P N].

The tree is not interpretable because PRO has no meaning and no type. Therefore, PRO has to be QR-ed (PRO-movement) creating a λ-bound trace. This creates the interpretable tree:

(36) The LF

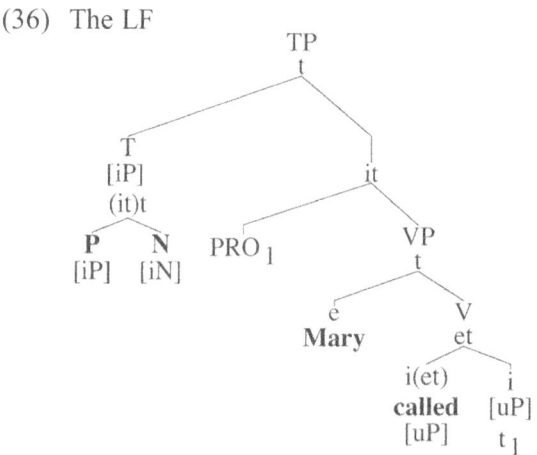

PRO_1 is the λ-operator $λ_1$. By means of this operator, [P N] transmits the feature [uP] to the bound variable t_1. [uP] agrees with the internal feature [uP] of **called**. Thus the finite verb always makes visible the semantic Tense of the sentence. Wolfgang Klein addresses this phenomenon as FIN-time in recent papers (see, e.g., Klein 2000).

A note on the literature. This system of feature transmission is very similar to that in Kratzer (1998). Kratzer claims that PRO (her zero tense $Ø_i$) has no temporal feature in the lexicon. The features are copied from the antecedent under binding. In von Stechow (2003) I claimed that PRO comes into being with a temporal feature because it is needed at PF for the morphology. The present approach is neutral to the question. PRO might enter the syntax without or with a temporal feature. The essential point is that the feature is checked under binding via agreement, presumably at LF. After checking the features may be deleted in agreement with Chomsky's views of the behaviour of features.

7. Temporal adverbials

In this section we consider the scope interaction of frame setting temporal adverbials like the one in the following sentence.

(37) Mary called on my birthday.

The simplest account of these temporal adverbials/PPs has them of type it. The adverbs are combined with the VP by Predicate Modification.

(38) [[**on my birthday**]] = $\lambda t.t$ is on my birthday

The following is the analysis we find in Heim (1997).

(39) Mary called on my birthday.
[P N] λ_1 [[Mary called t_1] [t_1 on my birthday]]
= $(\exists t_1)[t_1 < s^*$ & t_1 is on my birthday & Mary calls at t_1]

Here are the relevant lexical entries:

(40) Temporal Ps, type i(it)
 a. F(**on/in**) = $\lambda t.\lambda t'.t' \subseteq t$
 b. F(**at**) = $\lambda t.\lambda t'.\lambda P_{it}.t' = t$

on and **in** mean the same, but **on** is restricted to days, weekends and perhaps few other dates. **in** is reserved to larger time spans (e.g. *in 1975*). In many cases, **on** is covert, for instance with deictic dates such as **yesterday, today, tomorrow**.

(41) Deictic dates: type i
F(**yesterday**) = the day before the day that contains s*.

So the temporal adverb *yesterday* is represented as **ON yesterday** with covert **on**. I am using capitals for phonetically covert material. This semantics explains at least partially what Klein (2000) called the Present Perfect Puzzle, i.e. the fact that the present perfect doesn't combine with adverbs that denote a past time.

(42) *John has called yesterday.
N λt_1 **has**(t_1) λ_2 [[$_{PartP}$ **John called**(t_2)] [t_2 on **yesterday**]]
= $(\exists t)$[the final point of t is s* & John calls *in* t & t $\subseteq$ yesterday]

ON yesterday modifies the XN-interval, which gives rise to an inconsistency because an interval that contains the speech time cannot be on yesterday. Note that the participle phrase **John called** has to be related to the perfect time by means of the Perfective operator **PF**, which is defined in (50). The analysis ignores this, but the truth-condition takes care of this by locating John's call *in* the XN. The logical syntax has to make sure that temporal adverbs cannot be combined with VPs that are below PF. Note that the feature [uN] of **has** makes sure that the XN-auxiliary is always embedded under the semantic Present.

Time names like **six o'clock** are embedded under the preposition **at**, which expresses identity. As Klein (1994) and many other authors have noticed, temporal PPs may create an ambiguity when they scopally interact with the perfect auxiliary:

(43) Mary had left at six.
   a. **[P N] λt [[have (t) λt Mary left(t)] [t at six]]**
      (∃t < s*) t at 6 o'clock & (∃t' < t) Mary leaves at t'
   b. **[P N] λt [have (t)] λt [Mary left(t) [t at six]]**
      (∃t < s*)(∃t' < t) t' at 6 o'clock & Mary leaves at t'

(43a) is a Past-time (reference) time modification: the leaving is before six. (43b) is a Past-Past-time (event time) modification, the leaving is at six.

An interesting problem, which was discovered by Ogihara (1994), arises when the object of a temporal preposition is a quantifier:

(44) John worked on every Sunday.
   a. **P N λt every Sunday λt' t on t' John worked(t)**
      (∃t < s*)(∀t')[Sunday(t') → t on t' & John works at t]
   b. **every Sunday λt'P N λt t on t' John worked(t)**
      (∀t') [Sunday(t') → (∃t < s*) [t on t' & John works at t]]

The analysis assumes that **every Sunday** is a temporal quantifier of type (it)t and means $\lambda P_{it}.(\forall t)[Sunday(t) \rightarrow P(t)]$. Neither of the two readings is correct. Reading (44a) entails that there is a past time that is on every Sunday. Reading (44b) entails that every Sunday contains a time before the speech time. What we want is the following reading:

(45) (∀t')[Sunday(t') & <u>t' < s*</u> → (∃t < s*) [t on t' & John works at t]]

In other words, the Sundays we quantify over must be restricted to the past. Where does this restriction come from? We know that quantifiers come with domain restrictions. Assuming a framework in the style of von Fintel (1994), we assume that the temporal determiner **every** takes a variable C of type it as its first argument. The proper representation of *every Sunday* would then be **every**$_C$ **Sunday** with the meaning $\lambda P.(\forall t)[\text{Sunday}(t)\ \&\ g(C)(t) \to P(t)]$. Assume that $g(C) = \{t \mid t < s^*\}$. Take now the LF in (44b) with the so restricted determiner. This will give you the meaning in (45).

8. Tenses as referring terms?

The semantics of the Past assumed here is indefinite, i.e. **P** is an existential quantifier, which shifts the local evaluation time backwards. Partee (1973) argued that such a theory cannot be correct. Tenses are more like pronouns. They refer to definite times. Her crucial example is this:

(46) I didn't turn off the stove.

$\neq$ a. $(\exists t < s^*)\ \neg$I turn off the stove at t
$\neq$ b. $\neg(\exists t < s^*)$ I turn off the stove at t

The first interpretation is obtained from an LF in which Past (i.e. **P N**) has wide scope with respect to the negation $\neg$. The second reading is expressed by an LF in which the negation out-scopes the semantic Past. The first reading is trivial and the second is too strong. Partee's conclusion is that the semantic Past must be something like a pronoun that refers to a past time the speaker has in mind. Partee herself did not present a formal treatment. A reasonable way to implement the idea is found in Heim (1994b):

(47) Referential Past: type ii.
$[[\text{PAST}]]^{s^*} = \lambda t: t < s^*.t$

PAST is a function that takes a time as an argument and is only defined if the time is before s* If this is so, the value of the function is that time. The argument of PAST is given by a temporal variable. It is tempting to symbolize Partee's sentence as:

(48) NOT I turn off (PAST(t_5)) the stove
$= \neg$I turn off the stove at $g(t_5)$, where $g(t_5) < s^*$

t_5 would then denote a particular time in the past, which the speaker has in mind. As realized by Partee herself later on this cannot be the whole story, however. The action of turning off the stove takes a very short time, and it is virtually impossible that the speaker can refer to that time. The speaker might have in mind a longer time stretch, say the time from 11 to 12 AM, the period before her leaving. But this means that the meaning of **turn_off** cannot be the function [λt.λx.λy.y turns off x <u>at</u> t]. It rather has to be [λt.λx.λy.y turns off x <u>in</u> t]. But what does it mean that the action of turning off the stove occurs <u>in</u> the interval t? It means that it occurs *at some subinterval of t*. So a lexical entry of the verb that is compatible with Partee's account must be this:

(49) An accomplishment for Partee
    [[**turn_off**]] = λt.λx.λy.(∃t')[t' ⊆ t & y turns off x <u>at</u> t]

While this gives the correct truth-condition for Partee's sentence, this is presumably not a prosperous route to go down. To mention one difficulty: it seems hard to give a semantics for the Progressive **turning off** on the basis of this lexical semantics. In fact, this semantics would better be decomposed into an **at**-semantics for the verb plus a semantic Perfective operator, i.e. an operation that localises the event time within some other time (normally the "reference time" or "topic time").

(50) The Perfective, type t((it)t) (Krifka 1986; Klein 1994)
    [[**PF**]] = λt.λP$_{it}$.(∃t') t' ⊆ t & P(t')

The Partee sentence would then better be analysed as:

(51) **PAST(t_5) λt ¬ PF (t) λt I turn_off(t) the stove**
    = ¬(∃t ⊆ t_5) I turn off the stove at t, where t_5 < s*

An approach of this kind has become rather popular (see e.g. Abusch 1994; Heim 1994b; Kratzer 1998), but a closer inspection shows that it undermines the arguments of Partee against an indefinite analysis of time. By λ-conversion, the LF in (51) is equivalent with

(52) **¬ PF(PAST(t_5)) λt I turn_off(t) the stove**

If we can make sure that the indefinite Past tense means precisely **PF(PAST(t_5))**, i.e. (∃t ⊆ t_5), then the indefinite tense theory is restored and the originally rejected LF (46b) proves to be the (basically) correct one.

9. Restricting Tense

Our indefinite Past is an existential quantifier and comes with a domain restriction. So the refined lexical entry is this:

(53) Contextually restricted Past: type (it, i(it,t))[9]
$[[\mathbf{P}]] = \lambda C. \lambda t.\lambda Q.(\exists t')[C(t') \& t' < s^* \& Q(t')]$

The analysis of the Partee sentence is then:

(54) **not [P$_C$ N] λt. I [turn-off(t) the stove]**
If $g(C) = \{t \mid t \subseteq [11 \text{ a.m.}, 12 \text{ a.m.}]\}$, the LF means:
$\neg(\exists t < s^*) t \subseteq [11 \text{ a.m.}, 12 \text{ a.m.}]$ & I turn off the stove at t

This is almost the same truth-condition as before. The only difference is that the past time quantified over is presupposed in the first account but introduced by an existential quantifier in the second approach. The second approach contains a "definite part" as well, viz. the times in C. The first approach contains an indefinite part as well, but it is contained in the lexical entry. Note that we don't need the Perfective operator to obtain the correct meaning.

The domain restriction is useful for other purposes as well. I has been noted by many scholars that the event time of a Future Perfect construction should be after the speech time:

(55) John will have left at six.

We obtain the correct result if we assume the following pragmatic principle:

(56) An embedded tense or temporal auxiliary adds the "content" of the next super ordinate tense to its restriction provided the super ordinate tense is compatible with the tense in question.

The "content" of a Past are the times before s*, the content of a future are the times after s*, and so on. We have to add domain variables for the lexical entries of temporal auxiliaries, of course. The analysis of (55) is then the following:

---

[9] Restrictions of this kind have been introduced in Musan (2002).

(57) N λt WILL$_{C1}$(t) λt t at 6 have$_{C2}$(t) λt john left(t)
   g(C1) = {t | t a time}
   g(C2) = {t | t > s*}
   = (∃t)[t > s* & t = 6 & (∃t')[t' > s* & t' < t & John leaves at t']]

10. Scope interactions with Tense

Negation and quantifiers interact scopally with Tense. As we have seen, the negation usually takes wide scope. This makes correct predictions for the following sentence:

(58) It didn't rain today.
   not P N λt rain(t) t ON today
   = ¬(∃t < s*) t ⊆ today & it rains at t

Note that it is impossible to analyse this sentence correctly by means of the original Partee approach, i.e., the approach without the Perfective operator. The best approximation to the meaning using the Perfective is the following:

(59) **PAST(t$_5$) λt ¬ PF(t) λt rain(t) t ON today**
   = t$_5$ ⊆ today & ¬(∃t ⊆ t$_5$) it rains at t, t$_5$ < s*

As it stands the reading is too weak, because t$_5$ might be very short. Intuitively, however, the sentence means that the time in today that precedes s* is without rain. In order to get that reading we need a pragmatic principle stating that the reference time has to be as long as possible. Musan (2002: ch. 3) assumes such a principle. Another feature of this analysis is that the temporal adverb **ON today** has to be in the scope of the Perfective operator. Our short discussion of the Present Perfect ended with the conclusion that this option should be excluded. The indefinite analysis in (58) covers the facts without additional assumptions and seems preferable.

As noticed by Heim (1997), quantifiers tend to take wide scope with respect to semantic tense as well. This accounts for some facts observed in Cresswell (1979):

(60) John polished every boot.

'John polish a boot' is an accomplishment that can only be predicated of an interval. The point of the example is that we need a different past interval for each boot because John cannot polish each boot at the same time. We obtain the desired reading by scoping **every boot** over the semantic Past:

(61) **every boot λx P N λt John polished(t) x**
= (∀x)[boot(x) → (∃t < s*)[John polishes x at t]]

Again, an analysis in Partee's original approach is not possible. A referential theory of Tense needs the Perfective operator in order to achieve temporal co-variation for the quantifier. So the indefinite semantics for the Past has some advantages over the referential theory.

11. Subordinate Tense

A touchstone for the adequacy of the semantics for tense is the behaviour of tenses in subordinate constructions. The literature discusses three different types of constructions: (i) tense in relative clauses; (ii) tense in complement clauses, i.e., tense under attitudes; (iii) tense in adjunct clauses, notably **before/after**-clauses. In languages like Russian or Japanese, the facts are relatively easy to analyse. Unfortunately, this is not so for English, where each of these constructions requires a different strategy.

11.1. Tense in relative clauses

The most interesting observation, due to Ogihara (1989), concerns the behaviour of relatives under Future.

(62) Mary will buy a fish that is alive.
    a. ST = MT   (simultaneous)
    b. ST = s*   (deictic)

(63) Mary will buy a fish that has been alive.
    ST < MT   (shifted)
    ST < s*   (deictic, Perfect)

(64) Mary will buy a fish that was alive.
    ST < s*   (deictic)
    ? ST < MT   (shifted).

I am using the following abbreviations: ST = the event time of the subordinate clause, MT = the event time of the matrix clause. The important observation is that the simultaneous reading in (62), i.e., the fish is alive at the time of the buying, which is in the future, is expressed by a present form. Since this event is in the future, the semantic tense correlated to this morphology cannot be a semantic Present because the latter denotes the speech time.

The facts of English can be described by the following assumption:

(65) The semantic tense of the relative clause is obligatorily bound, here by a higher tense.[10]

A semantic tense is bound if the semantic Present **N**, which is part of it, is replaced by a temporal pronoun **Tpro**, which is obligatorily bound. Details aside, this is the account given in Kusumoto (1999).[11]

Here are the analyses of sentence (62):

(66) N λ_1 will(t_1) λ_2 M. buy(t_2) a fish WH$_3$ Tpro$_2$ λ_4 is(t_4) λ_5 x$_3$ alive(t_5)
    iN   uN   uN   uN   uN
= ($\exists t > s^*$)($\exists x$)[fish(x) & alive(x,t) & buy(Mary,x,t)]

This is the simultaneous reading. **Tpro** is bound by **will** but it gets its uN-feature from the matrix **N** via a binding chain. This feature is ultimately realised by the finite verb **is**. Here we have the first non-trivial case of feature transmission under semantic binding. It is worth noticing that tenseless forms like the infinitive (or adjectives and the past participle) transmit temporal features under binding.

A note on the analysis of the relative clause. It is formed by moving the semantically empty relative pronoun WH to [Spec,C] thereby creating a λ-abstract, represented as [WH$_1$ α]. The relative clause is combined with the

---

[10] When a relative clause is in the scope of an attitude, the semantic tense may also be bound by the attitude predicate (see below).

[11] Kusumoto uses a distinguished temporal variable t* for our **N**. In matrix sentences, t* denotes s*. In subordinate contexts, t* may be bound. The temporal argument of the semantic Past is a variable **past**$_i$ which denotes only if g(**past**$_i$) < s*, the semantics of Kratzer (1998). Basically, this is the meaning of referential Past. Since this variable is always bound in Kusumoto's system, this complication doesn't buy anything, as far as I can see. Therefore, it is not adopted to the present approach.

head noun **fish** by Predicate Modification. The LF ignores the fact that the object "a fish that is alive" has to be QR-ed for type reasons. All this is standard.

The deictic reading of (62) requires binding of **Tpro** to the matrix **N**:

(67) **N** λ_1 **will(t$_1$)** λ_2 **M. buy(t$_2$) a fish WH$_3$ Tpro$_1$** λ_4 **is(t$_4$)** λ_5 **x$_3$ alive(t$_5$)**
= $(\exists t > s^*)(\exists x)[\text{fish}(x) \ \& \ \underline{\text{alive}(x,s^*)} \ \& \ \text{buy}(\text{Mary},x,t)]$

The semantics is different, but the feature transmission is as before. I leave it to the reader to provide the analyses for the sentences in (63) and (65).

The grammatical status of the shifted reading of (65) is doubtful. Abusch (1996) and Kratzer (1998) accept it. Their example is *Next month I will answer every e-mail that arrived*. The analysis of these sentences has a semantic Past of the form **P(Tpro$_1$)** in the relative clause, where **Tpro$_1$** is bound to the matrix **N**. If we want to block such constructions, we have to stipulate that the semantic tense in relative clauses is always an obligatorily bound **Tpro** and nothing else.

Our account of feature transmission makes the correct prediction that the Present in the following relative clause can only have a deictic interpretation.

(68) Mary talked to a boy who is crying.
deictic√, simultaneous*

As before we obtain the deictic reading by binding the **Tpro** to the matrix **N**. We could bind **Tpro** to the matrix Past, but then the transmitted temporal feature would be [uP], which would be in conflict with the inherent feature [uN] of the auxiliary **is**.

For Past under Past constructions, the literature assumes three readings:

(69) Mary talked to a boy who was crying.
    a. ST = NT   (simultaneous)
    b. ST < MT   (backward shifted)
    c. ST > MT   (independent, forward shifted)

The forward shifted reading is prominent in the following sentence, which is due to Ogihara:

(70) Hillary married a man who became the president.

The simultaneous reading is obtained by having a **Tpro** (and no **P**) in the relative clause, where **Tpro** is bound to the matrix Past. The backward shifted reading is obtained by a semantic Past in the relative, which is bound to the matrix Past, and the independent reading has an embedded Past bound to **N**. The latter reading is a deictic reading, and the forward shifting is purely pragmatic. Obviously, this interpretation could deal with the two remaining cases in (69) as well. So Past under Past doesn't give us convincing data that require bound tense in the relative clause. We could have a deictic Past in all these cases. But attitudes will give us data that can only be dealt with within the binding approach:

(71) John thought that he would buy a fish that <u>was</u> still alive.   (Ogihara)

The event time of the relative clause certainly isn't before the speech time. Thus a referential analysis is impossible. Let us look to tense under attitudes then.

11.2. Tense under attitudes

To deal with attitudes we have to enrich our semantic framework. For this purpose we introduce an intensional λ-language. "Intensional" means that expressions of type a express meanings of type (sa), i.e., a-intensions. Recall that s is the type of possible worlds W, D_{sa} is the set of (partial) functions from W into D_a. In particular, each lexical entry will denote an intension. So the former lexical entries have to be revised to take care of the world argument. Here are some examples:

(72) Some revised lexical entries:
   a. $F(\mathbf{John_e}) = \lambda w.\text{John}$
   b. $F(\mathbf{called}_{i(et)}) = \lambda w.\lambda t.\lambda x.x$ calls in w at t.
   c. $F(\mathbf{N_i}) = \lambda w.s^*$
   d. $F(\mathbf{P}_{i(it,t)}) = \lambda w.\lambda t.\lambda P_{it}.(\exists t' < t)P(t')$

The interpretation of the language $[[\,.\,]]^{w,g}$ is as before. The only innovation is Heim & Kratzer's rule Intensional Functional Application (IFA).

(73) Recursive definition of the interpretation function $[[\,.\,]]_{F,g}$
   1. Let α be a lexical entry of type a . Then $[[\alpha]]^{w,g} = F(\alpha)$.
   2. Let x be a variable of type a. Then $[[x]]^{w,g} = \lambda w.g(x)$, $g(x)$ in D_a.

3. **FA**: Let α have type b and daughters β of type ab and γ of type a.
   $[[\alpha]]^{w,g} = [[\beta]]^{w,g}(w)([[\gamma]]^{w,g}(w))$
4. **IFA**: Let α have type b and daughters β of type (sa)b and γ of type a.
   $[[\alpha]]^{w,g} = \lambda w.\ [[\beta]]^{w,g}(w)([[\gamma]]^{w,g})$
5. **PM**: Let α have type a and daughters β and γ of the same type.
   $[[\alpha]]^{w,g} = \lambda w.\lambda x.[[\beta]]^{w,g}(w)\ \&\ [[\gamma]]^{w,g}(w)$
6. **Abstraction**: Let x be a variable of type a and let α be an expression of type b.
   $[[\lambda x\ \alpha]]^{w,g} = \lambda w.\lambda u \in D_a.[[\alpha]]^{w,g[x/u]}(w)$

The standard way of defining complements of predicates of attitudes is to say that they are simply propositions, i.e. sets of worlds. For instance, the sentence *It is raining* (Progressive ignored) would mean [λw.it rains in w at s*]. A straight forward semantics of attitudes in the style of Hintikka (1969) would be this:

(74) *believe*, type (st)(i,et)

F(**believe**) = $\lambda w.\lambda p_{st}.\lambda t.\lambda x.(\forall w')$[w' is compatible with everything x believes of w in w at time t $\rightarrow$ p(w')]

John believes it is raining could then be analysed as:

(75) **N λt John believes(t) [N λt is(t) λt raining(t)]**   (to be revised)

= $\lambda w.\ (\forall w')$[w' is compatible with everything John believes of w in w at time s* $\rightarrow$ it is raining in w' at s*]

Since the sentential complement is of type t whereas **believes** requires a complement of type st, we have to use IFA for the semantic composition. The example embeds a tensed proposition with deictic Present. Now it has been known at least since Prior (1967) that tense under attitudes cannot have a deictic interpretation. The sentence *John believed that it was raining* it used to report a belief of John's that he would have worded as: "It <u>is</u> raining". In other words, an embedded Past is used to express a "subjective now". In order to express this, the embedded Tense has to be bound. Using the strategy we know from the interpretation of relative clauses, we may assume a **Tpro** in the embedded clause and bind it to the matrix Past:

(76) John believed it was raining (to be revised)

P N λ_1 John believed(t_1) Tpro$_1$ λ_2 was(t_2) λ_3 raining(t_3)

= $\lambda w.(\exists t_1 < s^*)$ John believes in w at t_1 $\lambda w'$.it is raining in w' at t_1

This analysis correctly accounts for the licensing of the embedded past tense morphology, because the matrix Past transmits its [uP]-feature to **was** via the binding chain. And the semantics looks reasonable. An anaphoric of this kind has been proposed by Gennari (2003) among others.

It has been know for a long time that this cannot be the whole story (von Stechow 1984, 1995; Abusch 1994; Heim 1994b; among others). The analysis assumes that the subject knows precisely the time at which he is, but we are mostly wrong about the time. Consider the following sentence still assuming the anaphoric approach:

(77) At 5 o' clock Mary thought it was 6 o'clock.

P N λ_1 t_1 at 5 o' clock Mary thought(t_1) Tpro$_1$ λ_2 was(t_2) λ_3 t_3 at 6 o'clock

= $\lambda w.(\exists t_1 < s^*)$ $t_1 = 5$ o'clock & Mary thinks in w at t_1 $\lambda w'.t_1 = 6$ o'clock.

According to this analysis, the content of Mary's belief is the proposition that 5 o'clock is 6 o'clock, a blatant contradiction. Intuitively, however, there is nothing wrong with Mary's belief, she simply believes that the time at which she is located is 6 o'clock.

A solution of the problem following Lewis (1979) is this: despite the morphological appearance, the complement of the attitude predicate is not a temporally independent proposition of type st but a temporally dependent proposition of type s(it), i.e. a property of times. In other words, the clausal complement of the sentence in (77) is the proposition that the time t_1 is 6 o' clock but the property of being at 6 o'clock. We obtain this property by abstracting **Tpro$_1$** away. The semantics of the verb of attitudes has to be revised accordingly.

(78) *believe*, type (s(it))(i,et)   (style of Lewis 1979)

F(**believe**) = $\lambda w.\lambda P_{s(it)}.\lambda t.\lambda y.(\forall w')(\forall t')[(w',t')$ is compatible with everything y believes of (w,t) in w at time t $\to$ p(w')(t')]

((w,t) may be thought as that part of the world history w that is at time t. The antecedent of the conditional is abbreviated as (w',t') $\in$ Dox$_y$(w,t).) The revised LF for the sentence in (77) is the following:

(79) **P N λ_1 t_1 at 5 o'clock Mary thought(t_1) ~~PRO~~ λ_4 t_4 λ_2 was(t_2) λ_3 t_3 at 6 o'clock**

   iP   uP   uP   uP

= $\lambda w.(\exists t_1 < s^*)[t_1 = 5$ o'clock & $((\forall w',t') \in$ Dox$_{\text{Mary}}$(w,t_1)) t' = 6 o'clock]

In other words, Mary locates her time at 6 o'clock, and she does that at 5 o'clock.

The LF for the complement is created by starting with a temporal PRO at the Tense position. Since PRO is semantically void, it has to be moved for type reasons and creates a complement of the correct type. This is precisely the analysis proposed by Kratzer (1998). If we look at the LF, we see that the verb of attitude now qualifies as a binder of the embedded temporal variable t_4 and therefore also of the temporal variable of **was**. Thus there is a binding chain form the matrix Past up to **was**.

A semantic Past under an attitude can have a shifted reading. We obtain this by assuming a locally bound **P** in the complement:

(80) Mary thought Bill left.   (shifted)

P N λ_1 Mary thought(t_1) λ_2 P(t_2) λ_3 Bill left(t_3)

= $\lambda w.(\exists t_1 < s^*)$ Mary thinks in w at t_1 [$\lambda w'.\lambda t_2.(\exists t_3 < t_2)$ Bill leaves in w' at t_3]

The temporal behaviour of complement clauses can be described as follows:

(81) Tense in clausal complements

The semantic tense of a complement is either the semantically empty **PRO** or **P(PRO)**. **PRO** has to be moved for type reasons and thus creates a temporal abstract.

We are now in a position to deal with Ogihara's sentence (71):

(82) John thought that he would buy a fish that <u>was</u> still alive.   (Ogihara)

P N λ_0 J. thought(t_0) λ_1 t_1 λ_2 would(t_2) λ_3 he buy(t_3)
iP   uP   uP   uP   uP

a fish WH$_4$ was(t_5) λ_6 x$_4$ alive(t_6)
     Tpro$_3$ λ_5   uP

= $\lambda w.(\exists t_0 < s^*)$ J. thinks(w,t_0,[$\lambda w'.\lambda t_1.(\exists t_3 > t_1)(\exists x)$[fish(x,w') & alive(x,w',t_3) & buy(J.,x,w',t_3)]])

Note that the **Tpro$_3$** in the relative clause is bound buy the future **would**, the past form of **will**. This accounts for the fact that the temporal variable t_5 of **was** denotes a time after the subjective now of John. This is the crucial fact for the theory of tenses in relative clauses: we said that the semantic tense in relative clauses is a **Tpro** that is obligatorily bound. While the previous examples of a past tense in relative clauses could be treated as deictic tenses, this example has to be treated as a bound tense.

Notes on the literature: Temporal **PRO** is what Kratzer (1998) calls a zero tense. She writes it as Ø and the variable created by movement as $Ø_i$. For reasons of uniformity, Kratzer wants to have all tenses as pronouns, i.e. she assumes a referential semantics for Past (the one given in footnote 11). We have seen that an embedded Past can have a shifted reading. To get this, Kratzer assumes that the past tense is ambiguous between a referential Past and a Perfect, i.e. our **P** or **have**. Thus an indefinite analysis of the Past is needed in this theory as well. Ogihara (1989) deals with Tense under attitudes by a rule of tense deletion:

(83) Tense deletion in SOT languages (SOT-rule)

If Past/Present occurs in the scope of another Past/Present respectively, you may delete it.

(A tense is in the scope of another tense if the former is c-commanded by the latter.) Under attitudes the application of the rule is obligatory because the logical type of the attitude verb requires so. The result is roughly the same as in the present approach. The difference is that there is no trace of the embedded Tense in the construction. Apart from this, the LFs are alike. We will need the SOT rule in the next section.

11.3. Tense in temporal adverbial clauses

In this section we will be concerned with the temporal behaviour of *before/after*-clauses. Again, the most interesting data come from future constructions. Ogihara (1996: 5.5) quoting Stump (1985) provides the following paradigm.

(84) *after/before* under Future
   a. John will enter the room before Mary leaves.
   b. John will enter the room after Mary has left.
   c. John will enter the room after Mary leaves.
   d. *John will enter the room after Mary will leave.
   e. *John will enter the room before Mary will leave.

If the main clause contains a Past, we invariably find a Past in the adjunct clause.

(85) Mary left before/after John arrived.

Let as first ask what *before* and *after* mean. Sentences like the following suggest that the prepositions simply denote the relation of temporal precedence (<) and succession (>) respectively.

(86) Mary arrived after/before 6.

Sticking again to the initial extensional framework, the semantics is simply this:

(87) *before/after*   type i(it)
    F(**before/after**) = λt.λt'.t' </> t

The LFs of the sentences in (86) are therefore:

(88) **[P N] λt t after/before 6 λt Mary arrived(t)**
    = (∃t)[t < s* & t >/< 6 o'clock & Mary arrives at t]

The temporal adverbial is combined with the VP via Predicate Modification.
  Next consider the temporal adjunct clauses in (86). Heim (1997) proposes that the construction is analysed along the lines of the following paraphrase:

(89) Mary left before/after the time at which John arrived.

Here the complement is of type i, which is appropriate for the object position of the temporal preposition. Now this paraphrase is not sufficient because John might have arrived at many times. We therefore follow Beaver and Condoravdi (2003) and proceed along the following paraphrase:

(90) Mary left before/after the earliest time at which John arrived.

"time at which John arrived" is a relative clause of type it. (With *after* the operator "the latest" might be more appropriate in many cases. The analysis would be entirely parallel to the following.) The information "the earliest" is provided by the following coercion operator:

(91) EARLIEST: type (it)i
    [[EARLIEST]] = λP_{it}.the earliest time t such that P(t).
            = the t, such that P(t) & (∀t')[P(t') → t < t']

The analysis of (90) is this:

(92) **P N λ_1 [[Mary left(t_1)] [t_1 before EARL WH$_2$ P N λ_3 t_3 AT t_2 John arrived(t_3)]]**

= ($\exists t_1 < s^*$) M. leaves at t_1 & $t_1 <$ the earliest t_2 s.t. ($\exists t_3 < s^*$) $t_3 = t_2$ & J. arrives at t_3

= ($\exists t_1 < s^*$) M. leaves at t_1 & $t_1 <$ the earliest $t_2 < s^*$ s.t. J. arrives at t_2

In addition to the EARLIEST-operator the relative construction is covert, which you may not like. But the construction is semantically transparent and involves no new principles of composition.[12] In (92) we assumed a deictic tense in the complement. This not possible in the case Future constructions. We therefore apply the SOT-rule that has been introduced in the last section, i.e., we delete Present under Present:

(93) The interpretation of **before/after** clauses requires the application of the SOT-rule, i.e. dense deletion under c-command.[13]

A plain deletion won't do because the Tense occurs directly under a WH-operator, which requires a sentence of type t. After tense deletion, we have something of type it, and the WH-operator would convert it into something of type i(it), not a good argument for the EARLIEST-operator. Therefore we replace the Tense by a covert existential quantifier ∃ of type (it)t, which means $\lambda P_{it}.(\exists t)P(t)$. Here is the analysis of (84a):

---

[12] A further motivation for this analysis is that it provides a straightforward account of the sentences observed in Geis (1970). The sentence *I saw Mary in New York before she claimed that she would arrive* has two readings depending on the origin of the temporal relative pronoun WH. The LFs for the temporal adjuncts are these: (i) **t before EARL WH$_1$ ∃ λ_2 t_2 AT t_1 she claimed(t_2) λ_3 would(t_3) λ_4 she arrive(t_4)**; (ii) **t before EARL WH$_1$ ∃ λ_2 she claimed(t_2) λ_3 would(t_3) λ_4 t_4 AT t_1 she arrive(t_4)**. The interesting and intuitively prevailing reading is that with the second adjunct: Mary claimed that she would arrive on Sunday, but I met her on Saturday.

[13] Note that auxiliaries don't qualify as semantic tenses: **Have** is never deleted. Note further that we have to license the present of the embedded verb before the application of Tense Deletion.

(94) N λ_1 will(t_1) λ_2 John enter(t_2) t_2 before/after EARL WH$_3$ N $\exists$ λ_4 t_4 AT t_3 Mary leaves(t_4)

= ($\exists t_2 > s^*$) John enters the room at t_2 & $t_2 >/<$ the earliest t_3 s.t. Mary leaves at t_3.

This is the correct reading, provided the **EARLIEST**-operator quantifies over future times. To achieve this, we have to assume a domain restriction **C** for the operator that copies the semantic content of the next higher Tense, here **will**. After this refinement, we can explain the oddness of the sentences (84d/e): the earliest future time at which Mary *will* leave is a time right after the speech time. This creates nonsense or at least confusion.

To summarize this section: we can analyse the facts by means of Ogihara's SOT-rule with successive insertion of an existential quantifier. The result is a configuration that is different from the two preceding ones: the construction is not anaphoric nor does it contain a zero tense. The evaluation of the construction shows that the temporal variable of the verb is ultimately bound be the EARLIEST-operator.

Here are some remarks on the literature. The standard treatment of **after/before**-clauses is from Anscombe (1964). Anscombe treats the prepositions as generalised quantifiers mapping two temporal properties to a truth-value. There is no uniform semantics. **after** means $\lambda P_{it}.\lambda Q_{it}.(\exists t)(\exists t')[P(t) \& Q(t') \& t > t']$, and **before** means $\lambda P_{it}.\lambda Q_{it}.(\exists t)[P(t) \& (\forall t')[Q(t') \rightarrow t < t']$. Among other things, this approach has the problem of integrating semantic tense, a question Anscombe is not concerned with. Many authors, e.g. Heinämäki (1974), Ogihara (1996), Kusumoto (1999) use variants of this semantics and either have no semantic tense or admit that they face difficulty in integrating it. The first approaches that use the simple unified semantics of the prepositions are von Stechow (2002) and Beaver and Condoravdi (2003). Most ingredients in the present account are found in Heim (1997).

12. Conlusion

English has the Tenses, Present (**N**), Past (**P**), and the temporal pronouns **Tpro** and **PRO**. In addition we have the temporal auxiliaries **have** and **will**. The temporal morphology of finite verbs is licensed under semantic binding, which requires an open quantification over time in the syntax, i.e., verbs must have temporal variables. The treatment of tense in subordinate constructions requires different strategies. (i) Tense in relative clauses can be treated anaphorically; the Tense in the relative clause is an obligatorily bound **Tpro**. (ii) The Tense in complements of attitudes is PRO, which is moved for type reasons and creates a λ-abstract, i.e. complements of attitudes are tenseless. (iii) **after**- and **before**-clauses are treated either deictically or as tenseless; the latter requires a slightly different strategy from the treatment of attitudes: we delete a tense and replace it by $\exists_{i(it)}$. In this article I have treated some non-trivial constructions of English but by no means all the constructions that I am aware of. I am convinced that the system is general enough to be able to be applied to other constructions as well.

Appendix

Here is the most trivial example to illustrate how the truth-values of LFs are computed. The calculation is given to convince the reader that the system is entirely formal.

$[[\,[_{TP}\,[_T\,\textbf{P N}]\,\text{PRO}\,\lambda_1\,[_{VP}\,\textbf{John [called t}_1\textbf{]}]]\,]]^{\,g}$	LF for *John called*
iff $[[\,\textbf{P N}\,]]^{\,g}([[\lambda_1\,[_{VP}\,\textbf{John [called t}_1\textbf{]}]\,]]^{\,g})$	FA
iff $(\exists t < s^*)\,[[\lambda_1\,[_{VP}\,\textbf{John [called t}_1\textbf{]}]\,]]^{\,g}\,(t)$	FA and Lexicon
iff $(\exists t < s^*)\,\lambda t'.[[\,[_{VP}\,\textbf{John [called t}_1\textbf{]}]\,]]^{\,g[1/t']}\,(t)$	Abstraction
iff $(\exists t < s^*)\,[[\,[_{VP}\,\textbf{John [called t}_1\textbf{]}]\,]]^{\,g[1/t]}$	λ-conversion
iff $(\exists t < s^*)\,F(\textbf{called})(g[1/t](\textbf{t}_1))(F(\textbf{John}))$	FA and Lexicon
iff $(\exists t < s^*)\,F(\textbf{called})(t)(F(\textbf{John}))$	Definition of $g[1/t](\textbf{t}_1)$
iff $(\exists t < s^*)$ John calls at t	FA, Lexicon, λ-conversion

The last line stands for the truth-value 1. More complicated examples are treated analogously and may require very long computations.

Acknowledgement

I wish to thank Leah Roberts who corrected my English.

References

Abusch, Dorit
   1994   Sequence of Tense Revisited: Two Semantic Accounts of Tense in Intensional Contexts. In *Ellipsis, Tense and Questions*, Hans Kamp (ed.), 87–139. Stuttgart: Dyana-2 Esprit Basic research Project 6852.

> 1996 The now-parameter in future contexts. In *Context Dependency in the Analysis of Linguistic Meaning*, B. Partee and H. Kamp (eds.), 1–20. Stuttgart: IMS Working Papers.

Anscombe, G. Elizabeth M.
> 1964 Before and after. *The Philosophical Review* 74: 3–24.

Beaver, David and Cleo Condoravdi
> 2003 A Uniform Analysis of "before" and "after". Paper presented at *Proceedings of SALT 13*, Cornell, Ithaca.

Cresswell, Max J.
> 1979 Interval Semantics for some Event Expressions. In *Semantics from Different Points of View*, Rainer Bäuerle, Urs Egli and Arnim von Stechow (eds.), 143–171. Berlin: Springer.

Dowty, David
> 1979 *Word Meaning and Montague Grammar: The Semantics of Verbs and Times in Generative Semantics and in Montague's PTQ*: Synthese Language Library. Dordrecht: Reidel.

Geis, Michael
> 1970 Adverbial Subordinate Clauses in English. Ph.D. dissertation, MIT.

Gennari, Sivlia P.
> 2003 Tense Meanings and Temporal Interpretation. *Journal of Semantics* 20: 35–72.

Heim, Irene
> 1994a Puzzling reflexive pronouns in de se reports: Handout from Bielefeld conference.
> 1994b Comments on Abusch's theory of tense. Ms., MIT.
> 1997 Tense in compositional semantics: MIT lecture notes.
> 2005 Features on bound pronouns. Ms., Cambridge, MA.

Heim, Irene and Angelika Kratzer
> 1998 *Semantics in Generative Grammar*: Blackwell Textbooks in Linguistics. Oxford/Malden, MA: Blackwell.

Heinämäki, Orvokki Tellervo
> 1974 Semantics of English Temporal Connectives. Ph.D. dissertation, University of Texas at Austin.

Hintikka, Jaakko
> 1969 Semantics for Propositional Attitudes. In *Philosophical Logic*, J. W. Davis et al. (eds.), 21–45. Dordrecht: Reidel.

Kamp, Hans
> 1971 Formal Properties of "now". *Theoria* 37: 227–273.

Kamp, Hans and Uwe Reyle
> 1993 *From Discourse to Logic*. Dordrecht/London/Boston: Kluwer.

Klein, Wolfgang
> 1994 *Time in Language*. London/New York: Routledge.
> 2000 An analysis of the German perfect. *Language* 76(2): 358–382.

2002   The argument-time structure of recipient constructions in German. In *Issues in Formal German(ic)*, Werner Abraham and Jan-Wouter Zwart (eds.), 141–178. Amsterdam/Philadelphia: John Benjamins.
2008   Concepts of Time. *This volume.*
Kratzer, Angelika
1998   More Structural Analogies Between Pronouns and Tenses. In *SALT VIII*, D. Strolovitch and A. Lawson (eds.), 92–110. Cambridge, MA/Ithaca, NY: CLC-Publications.
Krifka, Manfred
1986   *Nominalreferenz und Zeitkonstruktion. Zur Semantik von Massentermen, Pluraltermen und Aspektklassen.* University of München.
1989   Nominal Reference, Temporal Constitution and Quantification in Event Semantics. In *Semantics and Contextual Expression*, R. Bartsch, J. van Benthem and van Emde Boas (eds.), 75–115. Dordrecht: Foris.
Kusumoto, Kiyomi
1999   Tense in embedded contexts. Ph.D. dissertation, University of Massachusetts at Amherst.
Lewis, David
1979   Attitudes De Dicto and De Se. *The Philosophical Review* 88: 513–543.
Montague, Richard
1973   The Proper Treatment of Quantification in English. In *Approaches to Natural Language. Proceedings of the 1970 Stanford Workshop on Grammar and Semantics.* J. Hintikka, J. Moravcsik and P. Suppes (eds.), 221–242. Dordrecht: Reidel.
Musan, Renate
2002   *The German Perfect. Its Semantic Composition and its Interactions with Temporal Adverbials*: Studies in Linguistics and Philosophy. Dordrecht/Boston/London: Kluwer.
Ogihara, Toshiyuki
1989   Temporal Reference in English and Japanese: University of Texas at Austin.
1994   Adverbs of Quantification and Sequence of Tense Phenomena. In *Proceedings from Semantics and Linguistic Theory IV*, M. Harvey and L. Santelmann (eds.), 251–267. Ithaca, NY: CLC Publications.
1996   *Tense, Attitudes, and Scope.* Dordrecht: Kluwer.
Partee, Barbara
1973   Some Structural Analogies between Tenses and Pronouns in English. *Journal of Philosophy* 70: 601–609.
Prior, Arthur N.
1967   *Past, Present, and Future.* Oxford: Oxford University Press.
Stump, Gregory
1985   *The Semantic Variability of Absolute Constructions*: Synthese Language Library. Dordrecht: Reidel.

Tailor, Barry
   1977   Tense and Continuity. *Linguistcs and Philosophy* 1(2): 199–222.

Vendler, Zeno
   1957   Verbs and Times. *The Philosophical Review* 66: 143–160.

von Fintel, Kai
   1994   Restrictions on Quantifier Domains. Ph.D. dissertation, University of Massachusetts at Amherst.

von Stechow, Arnim
   1984   Structured Propositions and Essential Indexicals. In *Varieties of Formal Semantics. Proceedings of the 4th Amsterdam Colloquium, September 1982*, Fred Landman and Frank Feldman (eds.), 385–404. Dordrecht: Foris.
   1995   On the proper treatment of tense. In *SALT V*, Teresa Galloway and Mandy Simons (eds.), 362–386. (Cornell Working Papers in Linguistics.) Ithaca, NY: CLC Publications.
   2002   Temporal Prepositional Phrases with Quantifiers: Some Additions to Pratt and Francez (2001). *Linguistics and Philosophy* 25: 755–800.
   2003   Feature deletion under Semantic Binding: Tense, person, and mood under verbal quantifiers. In *NELS 33*, Makoto Kadowaki and Shigeto Kawahara (eds.), 397–403. Amherst, MA: GLSA.

Zeijlstra, Hedde
   2004   *Sentential Negation and Negative Concord*. Utrecht: LOT Publications.

Temporality in first and second language acquisition

Yasuhiro Shirai

1. Introduction

This chapter reviews the acquisition of temporal expressions in first and second language. The surveys of the systems of temporal expressions in various languages of the world in this volume make abundantly clear how complex the systems are that children and adult second language learners need to acquire to express temporality through language. How is it possible to acquire them? Although there is a long way to go before we can fully understand the mechanism of acquiring temporal expressions, there has been consistent progress in our understanding of how learners acquire temporal expressions in their various forms.

This chapter surveys the numerous studies that have been done to address this issue. I will survey the pattern of general development of how temporal expressions are acquired by children and adults by focusing on three areas of development: pragmatic, lexical, and grammatical. Grammatical development has received by far the most attention, and thus will be given detailed treatment.

The review presented here does not go into details of individual studies; nor do I try to make it complete in its coverage, since such detailed reviews are available elsewhere. Rather, I will present the general picture, and then focus on the current issues, which I believe will further the research in this important area of inquiry.

2. Pragmatic means

It has long been known that both in first and second language acquisition, language learners rely on pragmatic information to make temporal reference. In a strict sense, relying on pragmatic information may not itself be 'temporal expression', but this sort of implicit means of establishing temporal reference is very powerful.

In first language acquisition, for example, Clark (2003: 258) suggests that children, in describing sequences of events, prefer to present them in the order in which they occurred; that is, the order of mention normally follows the actual order in which those events occurred in the real world.

Second language learners likewise follow the sequence of events at the beginning of their language development. Extensive functional analyses of second language speech (Meisel 1987; Klein 1986; Schumann 1987; von Stutterheim and Klein 1987), which Bardovi-Harlig (2000) called 'meaning-oriented studies', have found that L2 learners use various pragmatic means (e.g. interlocutor scaffolding, calendric ordering) to establish temporal reference.

What distinguishes L1 and L2 acquisition at this stage, of course, is the completely different level of conceptual development. Adult L2 learners know how to establish temporal reference in their L1, and can rely on various means that work for both L1 and L2. For example, they can say:

In China, I work doctor. (=In China, I worked as a doctor.)

Without using any explicit means, they can establish the temporal reference, assuming that the other participant in this conversation knows that the speaker was in China before coming to where they are now.

3. Lexical means

There are of course limitations to pragmatic means of expressing temporal reference. If one needs to establish temporal relations beyond the order that can be pragmatically inferred, more explicit means need to be used. Lexical means such as temporal adverbials (e.g. *yesterday, tomorrow*) and time conjunction (e.g. *when, before*) explicitly signal temporal location of an event.

Weist (1986), in a review of development of tense and aspect across languages, reports that children acquiring various languages start using temporal adverbials such as *yesterday, tomorrow, today*, at various time points in development: 3;0 in Italian, 2;6 in French and Spanish, 2;6 in Polish, and 2;2 in Mandarin.[1] Interestingly, children start using grammatical means much earlier than this (to be discussed in the next section).

---

[1] It is interesting that Chinese children develop temporal adverbs early, which may indicate that they learn to rely on lexical means more heavily for establishing temporal reference than children learning languages with rich morphological

A recent study by Huang (2003) supports the observation that children prefer to use grammatical markers rather than lexical. She directly investigated how children and their mothers mark the shift to past time reference in the mother-child interaction (children aged 3;2 and 3;3), and compared that with the same mothers' speech to other adults. She found that both of child-to-adult-speech relied heavily on aspect markers (61% and 57% for each), while adult-to-adult speech did not (only 2% and 4%). For temporal adverbials, it was just the opposite: adult-adult speech relied heavily on adverbials (75% and 77%) while children used them infrequently (16% and 10%). This clearly suggests a strong preference for grammatical marking by children.

In another systematic study, Pawlak et al. (2006), analyzing longitudinal corpora of English and Polish acquisition, showed that grammatical means are acquired earlier than adverbials. Further, it has been shown that in comprehension studies, younger children are not sensitive to temporal information supplied by adverbs (Smith 1980; Valian 2006).

In contrast, in adult second language acquisition, it has been shown that learners move from a pragmatic stage to a lexical stage by starting to use adverbials. Bardovi-Harlig (2000), by synthesizing European crosslinguistic research (e.g. Klein et al. 1993; Dietrich et al. 1995) suggests that adverbs of position (e.g. now, then), duration (all week) and frequency (twice, often) appear earlier than adverbs of contrast (already, yet).

4. Grammatical Means

In first language acquisition, children start using grammatical means (i.e. tense and aspect markers) to convey temporality quite early. Weist (1986) suggests that children start to use contrast using grammatical marking of tense-aspect at age 1;8 plus or minus 2 months. Thus, the use of lexical means of temporal reference is preceded by grammatical means by a wide margin.

This is in striking contrast to adult second language acquisition, where it has been established that learners go through three stages of development in temporal reference: pragmatic → lexical → grammatical (Bardovi-Harlig, 2000). In fact, the major crosslinguistic research funded by ESF (European Science Foundation, Klein and Perdue 1992; Perdue 1993) showed that this is not only the pattern observed for temporality, but also the general devel-

---

systems, since that is how a tenseless language like Chinese works (e.g. Smith and Erbaugh 2005; Wu 2002).

opmental pattern for second language, and in their longitudinal research on natural learning occurring in the target language community, it was shown that many untutored learners barely made the grammatical stage. Further, Bardovi-Harlig's (2000) research with instructed learners in an ESL program in the US showed that they too follow the same stages in the acquisition of temporality.

The difference between L1 and L2 can be attributed to two possibilities, although they are not mutually exclusive and both may be contributing. One is children's conceptual immaturity. Since their conception of time is still limited, they cannot properly map temporal notions to lexical items. The other is the difference between children and adults in their ability to process grammatical information.

Adult second language learners tend to rely on lexical information in language processing (e.g. VanPatten 2002). Since there is much redundancy in language, oftentimes adults do not have to process grammatical information. For example, hearing the sentence "Yesterday I walked to the school", one does not need to process the past tense inflection -*ed* for comprehension.

Children are known to process and acquire morphology very early. Newport (1990) proposed the "less is more" hypothesis, which suggests that children, due to their limited memory capacity, cannot pay attention to a larger span of linguistic information, which makes them more suited for local, feature-based learning, such as morphological marking.

Although the explanation for the adult-child differences in the preferred means of temporal expression is an intriguing issue, it is beyond the scope of this chapter, and we now move on to the major focus of this chapter (as well as of the field): The debate surrounding the acquisition of grammatical markers of temporality.

5. The acquisition of tense-aspect markers

The acquisition of grammatical means for temporal expression has received a great deal of attention in the literature. This attention to tense-aspect marking, however, does not necessarily concern its role as expression of temporality; rather, such attention has focused on the theoretical debates concerning morphology and syntax.

5.1. Past tense and the regular-irregular debate

Since the 1980's, the English past tense has been the center of attention as a result of Rumelhart and McClleland's (1986) connectionist model of past tense acquisition, which they claimed learned past tense rules without differentiating regular and irregular morphological systems. Such a network was argued to constitute an existence proof that linguistic rules were not necessary, but rather were emergent phenomena or approximations resulting from the pattern of connections and activations that drive our linguistic behavior. This met strong criticisms from the traditional symbolic camp (e.g. Pinker and Prince 1988) and the debate continues (e.g., Pinker and Ullman, 2002; McClleland and Patterson 2002; Kielar et al. 2008).

This interest in past tense marking, however, was not on the past tense as temporal expression to denote pastness – rather it happened to have a morphological organization that includes regular and irregular forms to denote the same meaning. The debate then extended to other domains with similar organization (e.g. plural marking, derivational morphology). There is an attempt to connect this issue to the issue of temporality (e.g. Shirai, in press), and future development is expected. This is important because children have to figure out both form (morphological organization) and meaning (temporality) together, even though researchers tend to take different areas apart.

5.2. Tense marking and syntax

Another line of research that has been pursued extensively was the acquisition of syntactic category in relation to tense. Since tense marking supposedly represents the presence of the corresponding syntactic category TENSE, researchers have attempted to identify when, how, and why it is acquired, or whether it is present right from the beginning. Particularly, the notion of root infinitive (RI, Wexler 1994; or optional infinitive, OI, Rizzi 1994) received much attention, based on the observation that early child language appears to lack finiteness (e.g. lacks past tense, third person singular present *-s* in English). Again, this interest largely focuses on the syntactic status of the forms rather than their nature as temporal expression.

However, recently this phenomenon has been pursued in relation to its temporal semantics – in particular, the role of lexical aspect. Gavruseva (2002, 2003) attempted to explain the lack of tense marking (and hence of the acquisition of the syntactic feature of tense) in L1 and L2 acquisition by resorting to lexical aspect. Hyams (2007) attempts to relate the issue of root infinitive to a larger picture of how children learn to interpret temporal reference in relation to the issue of semantic bias in the acquisition of tense-aspect markers. Here again, the field seems to be moving toward a more integrative picture of tense-aspect acquisition, a welcome trend that needs to continue.

5.3. Tense-aspect marking and verbal semantics

Since the 1970's, there have been interesting trends observed in the acquisition of tense and aspect marking. Children's use of past tense marking is not applied evenly to all past tense contexts, but appears to be restricted according to its temporal semantics. Brown (1973) and Bloom et al. (1980) in English, Bronckar and Sinclair (1973) in French, and Antinucci and Miller (1976) in Italian (and English) were the initial studies that were very influential. These studies show that when children use past tense marking (English simple past, French *passé composé*, Italian *passato prosimo*) they are not applied to all types of verbs but selectively attached to instantaneous, change of state verbs that denote clear end-results. On the other hand, progressive marking is only attached to activity verbs, and it is not incorrectly attached to stative verbs (Brown 1973). Thus from the beginning this semantic bias was shown to have a cross-linguistics basis, which was further extended to other languages: Turkish (Aksu 1978; Aksu-Koç 1989, 1998), Greek (Stephany 1981), Japanese (Shirai 1993), Chinese (Li 1990), Russian (Stoll 1998), Polish (Weist et al. 1984), German (Behrens 1993), and African American English (Green and Roeper 2007).

Such correlations have also been observed in second language acquisition, which has been extensively investigated crosslinguistically: Spanish (Andersen 1991), English (Robison 1991; Bardovi-Harlig and Reynolds 1995), French (Bardovi-Harlig and Bergström 1996), Italian (Giacalone Ramat 1995), Japanese (Shirai and Kurono 1998), and Korean (Lee and Kim 2007), among others.

Such associations between grammatical markers of tense-aspect and event types have received much attention; controversies continue both in first and second language acquisition, and several books on this topic have since been published (Li and Shirai 2000; Slabakova 2001; Salaberry and Shirai 2002; Ayoun and Salaberry 2007; Salaberry 2000; Rocca 2007).

Shirai (1991) proposed the following descriptive generalizations, which are now called the Aspect Hypothesis (Andersen and Shiai 1994; Shirai and Andersen 1995):

1. Learners first use past marking (e.g. English) or perfective marking (Chinese, Spanish, etc.) on achievement and accomplishment verbs, eventually extending its use to activities and stative verbs.
2. In languages that encode the perfective/imperfective distinction, imperfective past appears later than perfective past, and imperfective past marking begins with stative verbs and activity verbs, then extending to accomplishment and achievement verbs.
3. In languages that have progressive aspect, progressive marking begins with activity verbs, then extends to accomplishment or achievement verbs.
4. Progressive markings are not incorrectly overextended to stative verbs.

(Andersen and Shirai 1996: 533)

The semantic bias observed is also referred to as the Aspect-Before-Tense Hypothesis (Bloom et al. 1980), the Defective Tense Hypothesis (Weist et al. 1984), the Primacy of Aspect Hypothesis (Robison 1990), the Lexical Aspect Hypothesis (Salaberry 1999), the Aspect First Hypothesis (Wagner 2001).

In addition to first and second language acquisition, this line of research has been extended to other areas of language research: second language attrition (Yoshitomi 2007), sentence processing (Yap et al., in press), Specific Language Impairment (Stokes and Fletcher 2003; Leonard et al. 2007), and Alzheimer's Disease (Matthews 1989).

5.3.1. Issues in first language acquisition of tense-aspect marking

The semantic bias observed has generated controversies that are still not resolved. They include:

1. What is it that children's early tense-aspect marking encodes?
2. Why is it that children's tense-aspect marking shows semantic bias?

Although they are separate issues, they in fact are closely related because (1) involves a great deal of interpretation which is oftentimes theory-driven. Thus, although most agree on the existence of semantic bias in tense-aspect markers both in production and comprehension (Shirai, Slobin and Weist 1998),[2] there still is no agreement in the field about these two questions. Here I will organize the discussion according to the explanations that have been offered to account for the semantic bias (i.e., the second issue, and will come back to the first issue at the end.

5.3.2. Cognitive deficits

One explanation for the semantic bias that was proposed in the 1970's was cognitive deficits. That is, children are cognitively immature and restricted to the 'here and now', and therefore they cannot represent deictic past. As a result their past tense forms actually encode 'change of state' or 'result-state', and that is why their past tense marking is mostly restricted to telic verbs (accomplishments and achievements). Such a Piagetian explanation was offered by early studies such as Bronckart and Sinclair (1973) and Antinucci and Miller (1976).

However, this explanation was refuted by Weist et al. (1984) who showed that young Polish children do refer to remote past events at a very young age. Weist et al. convincingly argue that there is no such cognitive constraint that prevents children from using past tense form to refer to remote past.

But do cognitive limitations on the part of young children play no role for the lack of true deictic past use – e.g. past tense used with atelic verbs (states and activities)? In order to tease out the effect of verb semantics vis-a-vis cognitive limitation, Shirai (1991; see also Shirai and Andersen 1995) additionally coded for 'result-state' (i.e., whether the verb refers to an observable result in front of a child). It was found that the earliest past tense forms are not only [+telic, +punctual] (i.e. achievement) but also [+result]. This indicates that 'observable result' may also contribute to emergent past tense marking.

Thus, it appears that although a strong 'cognitive constraints' argument is not tenable, it is still likely that children tend to find it easier to give past marking to a resultative situation due to cognitive limitations. It is possible to use past tense for true deictic past, but it is easier still, especially at early stages, to give past tense marking for events that have observable results (see Shirai and Miyata 2006 for a detailed argument).

---

[2] It is true that most of the studies which reported such semantic bias use production data, both natural interaction data and experimentally elicited data. However, there are comprehension studies that also show the same semantic bias (e.g., Li and Bowerman 1998; Stoll 1998; Wagner 2001).

5.3.3. Aspect before tense

Another explanation for the semantic bias observed in tense-aspect marking relies on the notion of primacy of aspect over tense. Crosslinguistically, there appears to be some primacy of aspect over tense. More languages grammatically mark aspect than tense (Lyons 1977). When languages encode both aspect and tense grammatically, aspectual marking is placed closer to the verb stem than tense marking (Jakobson 1957; Bybee 1985).

These typological generalizations, however, concern grammatical aspect, which should be distinguished from lexical aspect. Thus, this principle can explain why past tense marking is associated with a particular lexical aspectual class (i.e. achievements and accomplishments), but cannot explain why progressive marking is associated with activity verbs, since the principle does not predict the primacy of lexical aspect over grammatical aspect.

5.3.4. Grammaticizable notions

Another proposal offered to explain the semantic bias is the idea that some semantic notions have privileged status and therefore receive grammatical marking. Slobin's (1985) Basic Child Grammar and Bickerton's (1981) Language Bioprogram Hypothesis are cases in point.

Slobin (1985) proposed that some semantic notions are salient to children, and that when children hear grammatical functors in the input, they try to map them onto such notions. In relation to tense marking, he proposed 're-sult' and 'process' as such notions, and that is why children crosslinguistically map past/perfective marking onto 'result' and progressive marking onto 'process'.

Bickerton (1981) proposed the Punctual-NonPuncutual Distinction (PNPD) Hypothesis as part of an innate bioprogram to account for the results from child Italian and French discussed above. Bickerton argued that children want to distinguish punctuals from non-punctuals (which is in fact the telic-atelic distinction at the level of lexical aspect) by using grammatical markers they hear in the input. Therefore, disregarding the adult grammar which marks past tense, children re-interpret the past tense marker as a marker of puncutality (i.e. telicity) and that is why past marking is associated with telic verbs. He also suggested that children acquiring English re-interpret progressive *-ing* as a marker of non-punctuality (i.e. lack of telicity).

Bickerton (1981) also proposed the State-Process Distinction (SPD) as an explanation for the observation that children do not make the error of attaching progressive marking to stative verbs (Brown 1973; Kuczaj 1978), which concerns generalization (4) of the Aspect Hypothesis. He argued that children do not overuse -*ing* to stative verbs (e.g. **I am knowing*...) because they are equipped with knowledge of the state-process (i.e. stative-dynamic) distinction.

These proposals[3] explain generalizations (1)–(4) of the Aspect Hypothesis quite well. However, studies done after the 1990's present results that are not compatible with these strong predictions – studies which we will turn to next.

5.3.5. Input-based prototype formation

It is known that there are strong associations between tense-aspect markers and lexical aspect of verbs (Comrie 1976: 72; Bybee 1985: 77). This is based on the naturalness of combination, according to Comrie. Some tense-aspect markers are more naturally attached with some situation types. For example, the duration of a punctual situation is so short that when we try to comment on it, it is almost always over, thus association with past tense form is frequent (Brown 1973; see Andersen and Shirai 1996 for a comprehensive review of the issues surrounding such skewing in the input).

Most of the previous accounts had not paid attention to such skewing observed in native speech, and therefore did not resort to it as a possible source for the semantic bias in learners' speech (and comprehension). Shirai (1991) directly investigated such distributional bias as an explanation for semantic bias in children's language. By analyzing caretaker-child interaction data from the CHILDES corpus (MacWhinney 2000), he found that about 60% of adults' use of progressive and past tense markings are attached to their prototypical use (activity verbs and achievement verbs, respectively), while in children's speech, the emergent inflections are mostly restricted to its prototype (about 95%) (see Shirai and Andersen 1995 for details) at the emergent stage of these morphemes in English.

Regarding generalization (4) of the Aspect Hypothesis, Shirai also found that the lack of overgeneralization of progressive marking to stative verbs is largely dependent on the input (Shirai 1994). He found that two mothers out of three did not use stative progressive at all, while one mother did, which in fact corresponded to the use of stative progressives by the children; i.e. the

---

[3] Both Slobin and Bickerton have revised their position since (see Slobin 1997, 2001; Bickerton 1999).

child of the mother who used stative progressives did use stative progressives frequently, including incorrect ones (e.g. *I seeing light).

This prototype hypothesis based on input frequency accounts for most of the observed data very well. In particular, it works very well with English data – about 60% skewing in the input results in almost 100% skewing in child language, which Shirai (1991) attributed to the formation of semantic prototypes on the part of the children. They actively reorganize information in the input to form a semantic prototype for each tense-aspect marker, namely [+result, +punctual, +telic, +past, +unitary] for past tense marking, and [–punctual, –telic, –past, +dynamic] for progressive marking (see Andersen and Shirai 1994). Li and Shirai (2000; see also Li 2000) implemented this model using a self-organizing neural network model (i.e. without negative evidence), and found results very similar to data observed in human children, thus supporting the feasibility of the input-based prototype model.

An apparent counterexample to this input-based prototype formation is Swift's (2004) work on the acquisition of Inuktitut. In this language, children start using past markers in non-resultative contexts with predominantly atelic verb stems in reference to past activities and states, thus completely going against the Aspect Hypothesis. This seemingly puzzling pattern of acquisition comes from the interesting organization of the Inuktitut tense-aspect system, in which zero-marked telic verbs (achievements and accomplishments) refer to past/perfective, zero-marked dynamic atelic verbs (activities) refer to ongoing situations, and zero-marked non-dynamic atelic verbs (states) refer to ongoing states. Therefore, for telic verbs, there is no need to give past marking – just zero form will do, and hence, past marking develops from atelic verbs. Although Swift did not look at input frequency, it is quite plausible that this early emergence of past tense form reflects what children hear in the input. If so, this language does not constitute a counterexample to the input-based prototype hypothesis.

5.3.6. Multiple-factor account

Distributional analysis and subsequent prototype formation can account for most of the results reported in the literature, but there are some problematic cases. Shirai (1998) found a large number of stative verbs attached to the emergent past form -ta in the acquisition of Japanese. Shirai explained that such deviation must be explained by resorting to multiple factors. In the case of Japanese children, their inflated use of statives with past tense was largely due to verbs of existence inflected for past -*ta* (e.g. *atta* or *ita*), which functionally corresponds to 'I found it', uttered when the child has found a missing entity through direct perception. This scene is very salient to children, which is why the form emerges very early.

Chen and Shirai (in press) also found a much higher proportion of -*le* attached to stative predicates than is predicted by the Aspect Hypothesis in the speech of children acquiring Chinese, which they attributed to saliency of such scenes and the lack of obligatory aspect marking that facilitates rote learning.

Note that Slobin's and Bickerton's grammaticizable notions cannot explain these data, since state verbs are not something learners should give perfective or simple past marking.

In sum, available crosslinguistic data suggest that children make semantic representations predicted by the Aspect Hypothesis, but the degree to which each language (or even each tense-aspect marker) does this is highly influenced by how the tense-aspect system is organized, and multiple factors, including input frequency, complexity of form-function mapping, saliency to children, and typological characteristics, determine the degree to which the data conform to the hypothesis.

5.3.7. The significance of the prototype formation

Leonard et al. (2007) provide important data concerning the difference between typically developing (TD) children and children with Specific Language Impairment (SLI). In a production experiment, they compared SLI children with both MLU-matched and age-matched TD children for the accuracy of production of regular past tense and past progressive. The following table is adapted from their Tables 1 and 2. The study looked at 16 children from each group, but the results here exclude children who scored 100% (because of a ceiling effect), although the trends were essentially the same in both analyses.

Table 1. Percentages producing the target item for children not performing at ceiling (adapted from Tables 1 and 2 from Leonard et al. 2007)

	SLI	TD-MLU	TD-Age
	(N=16)	(N=15)	(N=10)
prototypical -*ed*	37	83*	97*
non-prototypical -*ed*	33	69	85
	(N=14)	(N=10)	(N=8)
prototypical -*ing*	58	59*	81*
non-prototypical -*ing*	64	41	65

*significant effect of lexical aspect

What they found for TD children was exactly what the prototype hypothesis predicts: They were more successful in producing past tense with prototypical past tense verbs (punctual, telic, result), while being more successful producing progressive –*ing* with prototypical progressive verbs (durative, dynamic, atelic). What is interesting was that there was no significant effect of lexical aspect for SLI children.

We tend to see children's initial restriction to prototypes as a limitation – hence defective. However, these results show that TD children's ability to create semantic prototypes is an important ability for making generalizations concerning form-meaning relationships, which seems to be lacking in SLI children. The ability to create a semantic prototype as "a starting point" (Leonard et al. 2007) seems to be essential for efficient development of adult-like morphological representation.

5.3.8. *What does early tense-aspect marking encode?*

To answer this question is not easy, because, naturally, we cannot really tell what children are actually encoding in the grammatical marking in question. All we can do is infer what they are doing based on their overt behavior. One possible avenue that may advance our understanding of this issue is to investigate their neural behavior, though to my knowledge no such study has been done to address this issue. If, for example, children's processing of past tense forms is similar in neural activation to that of adults and there is no effect of lexical aspect, then we can be more confident that children are encoding tense, not aspect. But for now, there is no such study, and therefore we need to rely on behavioral studies. As a result, the interpretation is heavily influenced by one's theoretical orientation.

Children's linguistic behavior is erratic, and is influenced by numerous factors. Different studies use different approaches, and it is therefore not easy to identify the cause(s) of variation in the results. In this area of inquiry, the degree to which the use of tense-aspect marking adheres to the predictions of the Aspect Hypothesis is not uniform, some studies being strongly consistent with its predictions, others relatively deviating from it. The difficulty of arriving at a consensus on this issue is also compounded by crosslinguistic differences in the acquisition of tense-aspect marking. From the beginning, as noted above, the research in this area has had a crosslinguistic scope, with French, Italian, English, Turkish, Greek, Polish studied in the 1970's and the early 80's. And since cognitive or linguistic constraints proposed at the time were supposedly universal, variation across studies were not given prominence, the debate centering on whether tense-marking codes aspect, or alternatively, tense-aspect marking codes lexical aspect. Weist et al. (1984) criticized earlier work by characterizing it as "the Defective Tense Hypothesis", showing that Polish children can use past tense for atelic verbs as well as remote past situations, thus effectively falsifying the strong version of the aspect before tense hypothesis and the cognitive deficit hypothesis. Weist (1986, 1989, 2002) has since consistently argued that children use tense marking to encode tense (see also Pawlak et al. 2006).

Bloom and Harner (1989) and Andersen (1989) independently reanalyzed the data reported in Weist et al. (1984), and showed that perfective past tense markings are predominantly on telic verbs. Andersen (1989) argued that the absolute version of the Aspect Before Tense hypothesis does not hold, but that a relative version is still valid. Weist et al. (1984) were in fact aware that the most frequent categories are past perfective and achievement and accomplishment verb phrases. This means that even though they look at the same data, they come to different conclusions.

It appears that this difference largely comes from their theoretical orientations. As is well known in the field, there are two theoretical orientations that dominate language acquisition research, which Ingram (1989) terms Child Language and Language Acquisition. Bloom and Harner (1989: 208), in their response to Smith and Weist (1987), state:

> [t]he tension between these two perspectives has a long history (see Bloom 1970) ... The theoretical tension between formal and developmental theories of language acquisition was heightened with the statement by Wexler & Culicover (1980) to the effect that descriptions of child language data are not necessary for explaining language acquisition.

One of the assumptions dominant in formal linguistically oriented approaches is the continuity assumption – unless there is evidence to the contrary, one interprets child language as having the same grammar as adult language. Although Weist does not explicitly take a formal linguistic approach, his interpretation of data appears to be in line with this. Since there is no evidence to the contrary, past tense marking codes past, not grammatical or lexical aspect.

A recent contribution to this issue by Valian (2006) also comes from this theoretical orientation. Based on the comprehension experiments that showed that children are sensitive to the contrast involving *is/was* and *will/did* used as copula and auxiliary, Valian (2006: 251) argues, "From the beginning of combinatorial speech, children's grammars include a syntactic tense marker that is independent from aspect and includes the syntactic category verb".

Although she found that even two-year-old children show statistically significant sensitivity between past and nonpast forms, the actual results are not that clear cut. For example, the percentage of 2-year-olds who interpreted past form as non-past (in a non-adverb condition) is about 20% for the *will/do* aux contrast, which is quite impressive, but 60% for copula (*is/was* adjective) and 80% for progressive (*is/was* V-ing) (from her Figure 1). It is quite likely that children associate aux *did* as a lexical item with past events, since they often hear "Yes I did" with stress on *did*. (Valian's experiments gave stress to copula and auxiliaries for ease of comprehension.)

The point here is not to criticize Valian's methodology or interpretation but to highlight the different approaches taken by these researchers. Until there is evidence to the contrary, they assume children have the same grammar as adults, but it is not easy to determine what constitutes 'evidence' to the contrary; hence the debate among formalist researchers on whether early syntax is deficient (e.g. Wexler 1994).

Developmental theorists take a different view about children's grammar – they tend to look at children's language very closely to discover what children are doing with the grammatical marking in question (Bloom and Harner 1989; Ingram 1989). Thus, when we evaluate the debate concerning what is coded by children's tense-aspect marking, we need to be aware of the distinct assumptions made by researchers, which are not explicitly stated in their papers.

Shirai and Andersen's (1995) prototype account discussed earlier (see also Li and Shirai 2000) is a developmentalist proposal that offers to resolve the debate concerning whether children's past tense marking encodes (lexical) aspect or tense. Based on the observation that children's early past tense forms in English are restricted not only to past tense but also to achievement verbs with resultant states, they argued that children's past tense marking marks both tense and aspect, which corresponds to the "prototypical perfective" based on Dahl's (1985) typological survey of tense-aspect markers in languages of the world.

As noted above, this prototype account works very well with English, but not in some other languages such as Japanese (Shirai 1998), where past tense form is associated with stative verbs quite early. Thus, it appears that a more flexible approach is needed.

I suggest that the question of what children's early tense-aspect markers encode is a very complex one, and is not an issue that we can answer just by looking at children's behavior. It is clear that children's grammatical development is heavily influenced by the pattern observed in the input and the typological features of the language that they are exposed to. By trying to answer whether early child functors encode tense or aspect, we tend to lose the real picture of what they are doing with them, or how they move from their early system to adult-like representation. Rather, we should focus on investigating how children move from early restricted semantics to adult-like knowledge where they can consistently mark pastness (in the case of English).

5.3.9. Issues in second language acquisition of tense-aspect marking

Second language acquisition researchers have also extensively investigated the issue of semantic bias in tense-aspect markers. The obvious difference between L1 and L2 acquisition is the effect of L1. One important hypothesis advanced in relation to the Aspect Hypothesis is the Default Past Tense Hypothesis (Salaberry 1999, 2000, 2008), which is closely related to L1 effect.

Salaberry and Ayoun (2007: 20) state:

> The Default Past Tense Hypothesis predicts that during the first stages of L2 development, learners will attempt to mark past tense distinctions rather than aspectual distinctions, and in so doing will initially rely on a single marker of past tense, most typically the perfective form.

Empirical evidence for this hypothesis comes from Salaberry's work on L2 Spanish, which shows that learners' use of perfective past (preterit) is spread across various lexical aspectual classes at the beginning stage, not just restricted to telic verbs, while association with telicity increases as the learners' proficiency increases.

This deviation in Salaberry's (1999, 2000) results from the prediction of the Aspect Hypothesis appears to be the combination of three factors: (1) learners' L1 (English), (2) the classroom situation, (3) task (Shirai 1997). First, the learners' L1 (English) has a simple past tense marker, which is less aspectual compared with Spanish preterit (perfective past). Therefore, assuming transfer, learners at the beginning stage will show a lesser degree of association with telicity, since perfective past is more strongly associated with teilc verbs than is simple past. Second, learners in classroom settings tend to show more L1 transfer than those in naturalistic settings (Shirai, 1992). Finally, the task (film retell) allows learners to use their monitor in the sense that it is a past narrative and they know they should tell it in past tense.

Importantly, most of the studies Salaberry cites that are consistent with the Default Past Tense Hypothesis come from L1 English learners learning Romance languages (e.g. Hasbún 1995; Salaberry 2002) in classroom settings.

Bonilla (2008) reviewed previous research on L2 Spanish tense-aspect and concluded:

1. Studies that used a personal narrative task supported the Aspect Hypothesis.
2. The support for the Default Past Tense Hypothesis comes from studies that only used an impersonal narrative task.
3. Time-constrained tasks supported the Default Past Tense Hypothesis.
4. When participants were allowed planning time, results were consistent with the Aspect Hypothesis.

A comprehensive review of L2 Romance studies should clarify the conditions under which the deviation from the Aspect Hypothesis is obtained. In any event, although the L2 studies that investigated semantic bias in tense-aspect marking largely neglected the effect of L1, it appears to be stronger than previously expected.

L1 effect has recently received some attention, especially in the area of Japanese L2 acquisition of aspect. Nishi and Shirai (2007) and Nishi (2008), in investigating the acquisition of Japanese imperfective aspect marker *-teiru* by L1 English, Chinese, and Korean learners, have shown that L1 influence is much stronger than previously thought. In particular, Nishi (2008) argued that the pattern of acquisition that is consistent with the Aspect Hypothesis in Japanese may come from the fact that progressive is represented in all three languages. Sugaya and Shirai (2007) also found that in their oral task learners whose L1 has progressive marking find it easier to correctly use the progressive meaning of the imperfective *-teiru* than the resultative meaning. In a longitudinal, bi-directional study of child L2 acquisition of past time reference, Rocca (2002, 2007) also found the effect of L1 in that L1 Italian learners of English overused progressive marking to stative verbs, transferring the general imperfective past (*imperfetto*) from Italian, while English learners underused it.

Extending such findings, it may even be possible to argue that results from previous L2 research that supported the Aspect Hypothesis can also be partially attributed to L1 influence. Most of the L1s have corresponding past or perfective marking, for which the prototype is telic and punctual (cf. Dahl 1985), which may transfer to L2 learning of past or perfective marking. This may facilitate early acquisition of past tense with telic verbs. Likewise, acquisition of progressive marking with activities may be facilitated because in most languages, progressive marking marks action in progress that is obtained when imperfecive aspect marking is combined with activity verbs, but other meanings (such as habitual, futurate) are not always present (Bybee et al. 1994).[4]

Future research should systematically investigate the effect of the L1 by comparing different L1 groups acquiring the same language to tease out the effect of natural acquisitional processes from the effect of L1, as done by Nishi (2008). For example, what are the differences between L1 German learners (simple past) vs. L1 Spanish learners (perfective past) of English in their acquisition of English past tense? How about L1 German learners (no progressive) vs. Chinese learners (restrictive progressive, with action in progress meaning only) in the acquisition of the highly grammaticized, polysemous English progressive?

---

[4] Although not directly related to the Aspect Hypothesis, a series of studies by Slabakova and Montrul (e.g. Slabakova and Montrul 2002; Montrul and Slabakova 2003) also address the issue of L1 transfer in the acquisition of aspectual marking.

6. Conclusion

This chapter reviewed research on the acquisition of temporal expressions in L1 and L2 acquisition that has been done in the last 40 years or so. It presented an overall picture first, and then focused on issues surrounding the development of grammatical marking of tense and aspect that have generated a great deal of research and controversy.

In the future, we should try to integrate the enormous amount of research findings that have been accumulated in different theoretical and methodological orientations, and, as stated above, such attempts seem to be starting. In so doing, we should be aware of different assumptions that developmentalists and formalists take. As Bloom and Harner (1989: 208) put it, it is important to "make the theoretical tension between developmental and formal theorists explicit and to call for mutual respect between the two perspectives. Both make important contributions to our understanding of language acquisition."

Acknowledgments

I thank Courtney Andrew, Benjamin Friedline, Kevin Gregg and Wendy Martelle for their helpful comments on an earlier version of this chapter.

References

Aksu, Ayhan
    1978   Aspect and modality in the child's acquisition of the Turkish past tense. Ph.D. dissertation, Department of Psychology, University of California, Berkeley.

Aksu-Koç, Ayhan
    1988   *The Acquisition of Aspect and Modality: The Case of Past Reference in Turkish*. Cambridge, UK: Cambridge University Press.
    1998   The role of input vs universal predispositions in the emergence of tense-aspect morphology: Evidence from Turkish. *First Language* 18: 255–280.

Andersen, Roger W.
    1989   La adquisición de la morfología verbal. *Lingüística* 1: 89–141.
    1991   Developmental sequences: The emergence of aspect marking in second language acquisition. In *Crosscurrents in Second Language Acquisition and Linguistic Theories*, Thom Huebner and Charles A. Ferguson (eds.). Amsterdam/Philadelphia: John Benjamins.

Andersen, Roger W. and Yasuhiro Shirai
1994 Discourse motivations for some cognitive acquisition principles. *Studies in Second Language Acquisition* 16: 133–156.
1996 Primacy of aspect in first and second language acquisition: The pidgin/creole connection. In *Handbook of Second Language Acquisition*, William C. Ritchie and Tej K. Bhatia (eds.). San Diego, CA: Academic Press.

Antinucci, Francesco and Ruth Miller
1976 How children talk about what happened. *Journal of Child Language* 3: 169–89.

Ayoun, Dalila and M. Rafael Salaberry (eds.)
2007 *Tense and Aspect in Romance Languages*. Amsterdam/Philadelphia: John Benjamins.

Bardovi-Harlig, Kathleen
2000 *Tense and Aspect in Second Language Acquisition: Form, Meaning, and Use*. London: Blackwell.

Bardovi-Harlig, Kathleen and Anna Bergström
1996 The acquisition of tense and aspect in SLA and FLL: A study of learner narratives in English (SL) and French (FL). *Canadian Modern Language Review* 52: 308–330.

Bardovi-Harlig, Kathleen and Dudley W. Reynolds
1995 The role of lexical aspect in the acquisition of tense and aspect. *TESOL Quarterly* 29: 107–131.

Behrens, Heike
1993 The relationship between conceptual and linguistic development: The early encoding of past reference by German children. *Chicago Linguistic Society (Papers from the Parasession on Conceptual Representation)* 29: 63–75.

Bickerton, Derek
1981 *Roots of Language*. Ann Arbor, MI: Karoma.
1999 Creole languages, the Language Bioprogram Hypothesis, and language acquisition. In *Handbook of Child Language Acquisition*, William C. Ritchie and Tej K. Bhatia (eds.). San Diego, CA: Academic Press.

Bloom, Lois and Lorraine Harner
1989 On the developmental contour of child language: A reply to Smith and Weist. *Journal of Child Language* 16: 207–216.

Bloom, Lois, Karin Lifter and Jeremie Hafitz
1980 Semantics of verbs and the development of verb inflection in child language. *Language* 56: 386–412.

Bonilla, Carrie
2008 Review of task conditions in L2 Spanish acquisition of tense–aspect morphology Unpublished Ms., Department of Linguistics, University of Pittsburgh.

Bronckart, Jean-Paul and Hermine Sinclair
   1973   Time, tense and aspect. *Cognition* 2: 107–130.
Brown, Roger
   1973   *A First Language*. Cambridge, MA: Harvard University Press.
Bybee, Joan L.
   1985   *Morphology: A Study of the Relation between Meaning and Form.* Amsterdam/Philadelphia: John Benjamins.
Bybee, Joan L., Revere Perkins and William Pagliuca
   1994   *The Evolution of Grammar: Tense, Aspect, and Modality in the Languages of the World.* Chicago, IL: University of Chicago Press.
Chen, Jidong and Yasuhiro Shirai
   in press   The development of aspectual marking in child Mandarin Chinese. Submitted. *Applied Psycholinguistics.*
Clark, Eve V.
   2003   *First Language Acquisition.* Cambridge: Cambridge University Press.
Comrie, Bernard
   1976   *Aspect.* Cambridge, UK: Cambridge University Press.
Dahl, Östen
   1985   *Tense and Aspect Systems.* Oxford: Blackwell.
Dietrich, Rainer, Wolfgang Klein and Colette Noyau (eds.)
   1995   *The Acquisition of Temporality in a Second Language.* (Studies in Bilingualism 7.) Amsterdam/Philadelphia: John Benjamins.
Gavruseva, Elena
   2002   Is there primacy of aspect in child L2 English? *Bilingualism: Language and Cognition* 5: 109–130.
   2003   Aktionsart, aspect, and the acquisition of finiteness in early child grammar. *Linguistics* 41: 723–755.
Giacalone Ramat, Anna
   1995   Tense and aspect in learner Italian. In *Temporal Reference, Aspect and Actionality 2: Typological Perspectives*, Pier Marco Bertinetto, Valentina Bianchi, Östen Dahl and Mario Squartini (eds.). Torino: Rosenberg and Sellier.
Green, Lisa and Thomas Roeper
   2007   The acquisition path for tense-aspect: Remote past and habitual in child African American English. Language Acquisition 14: 269–313.
Hasbún, Leyla M.
   1995   The role of lexical aspect in the acquisition of the tense/aspect system in L2 Spanish. Ph.D. dissertation, Department of Linguistics, Indiana University.
Huang, Chiung-Chih
   2003   Mandarin temporality inference in child, maternal and adult speech on interpretation in early grammar. *First Language* 23: 147–169.

Hyams, Nina
    2007   Aspectual effects on interpretation in early grammar. *Language Acquisition* 14: 231–268.

Ingram, David
    1989   *First Language Acquisition: Method, Description, and Explanation.* Cambridge: Cambridge University Press.

Jakobson, Roman
    1957   Shifters, verbal categories, and the Russian verb. Cambridge, MA: Harvard University, Russian Language Project, Department of Slavic Languages and Literature.

Kielar, Aneta, Marc F. Joanisse and Mary L. Hare
    2008   Priming English past tense verbs: Rules or statistics? *Journal of Memory and Language* 58: 327–346.

Klein, Wolfgang
    1986   *Second Language Acquisition.* Cambridge: Cambridge University Press.

Klein, Wolfgang, Rainer Dietrich and Colette Noyau
    1993   The acquisition of temporality. In *Adult Language Acquisition: Cross-linguistic Perspectives: Vol. 2, The Results*, Clive Perdue (ed.). Amsterdam: John Benjamins.

Klein, Wolfgang and Clive Perdue (eds.)
    1992   *Utterance structure: Developing grammars again.* Amsterdam: John Benjamins.

Kuczaj, Stan A. II
    1978   Why do children fail to overgeneralize the progressive inflection? *Journal of Child Language* 5: 167–171.

Leonard, Laurence B., Patricia Deevy, Robert Kurtz, Laurie Krantz Chorev, Amanda Owen, Elgustus Polite, Diana Elam and Denise Finneran
    2007   Lexical aspect and the use of verb morphology by children with Specific Language Impairment. *Journal of Speech, Language, and Hearing Research* 50: 759–777.

Lee, EunHee and Hae-Young Kim
    2007   On crosslinguistic variations in imperfective aspect: The case of L2 Korean. *Language Learning* 57: 651–685.

Li, Ping
    1990   Aspect and aktionsart in child Mandarin. Ph.D. dissertation, Faculty of Letters, Leiden University, the Netherlands.
    2000   The acquisition of lexical and grammatical aspect in a self-organizing feature-map model. In *Proceedings of the 22nd Annual Conference of the Cognitive Science Society*, Lila Gleitman and Aravind Joshi (eds.). Mahwah, NJ: Lawrence Erlbaum.

Li, Ping and Melissa Bowerman
    1998   The acquisition of lexical and grammatical aspect in Chinese. *First Language* 18: 311–350.

Li, Ping and Yasuhiro Shirai
  2000   *The Acquisition of Lexical and Grammatical Aspect.* Berlin/New York: Mouton de Gruyter.
Lyons, John
  1977   *Semantics, Vol. 2.* Cambridge, UK: Cambridge University Press.
MacWhinney, Brian
  2000   *The CHILDES Project: Tools for Analyzing Talk.* Third Edition. Hillsdale, NJ: Lawrence Erlbaum.
Mathews, Stephen J.
  1989   Where was I...? Tracking tense and aspect in Alzheimer's disease. Unpublished manuscript, Department of Linguistics, University of Southern California.
McClelland, James L. and Karalyn Patterson
  2002   Rules or connections: What does the evidence rule out? *Trends in Cognitive Sciences* 6: 465–472.
Meisel, Jürgen M.
  1987   Reference to past events and actions in the development of natural second language acquisition. In *First and Second Language Acquisition*, Carol W. Pfaff (ed.), New York: Newbury House.
Montrul, Silvina and Roumyana Slabakova
  2003   Competence similarities between native and near-native speakers: An investigation of the preterite-imperfect contrast in Spanish. *Studies in Second Language Acquisition* 25: 351–398.
Newport, Elissa L.
  1990   Maturational constraints on language learning. *Cognitive Science* 14: 11–28.
Nishi, Yumiko
  2008   Verb learning and the acquisition of aspect: Rethinking the universality of lexical aspect and the significance of L1 transfer. Ph.D. dissertation, Linguistics, Cornell University, Ithaca, NY.
Nishi, Yumiko and Yasuhiro Shirai
  2007   Where L1 semantic transfer occurs: The significance of cross-linguistic variation in lexical aspect in the L2 acquisition of aspect. In *Diversity in language: Perspectives and implications*, Yoshiko Matsumoto, David Y. Oshima, Orin Robinson and Peter Sells (eds.). Stanford, CA: CSLI Publications.
Pawlak, Aleksandra, Jessica S. Oehlrich and Richard M. Weist
  2006   Reference time in child English and Polish. *First Language* 26: 281–297.
Perdue, Clive (ed.)
  1993   *Adult language acquisition: cross-linguistic perspectives: Vol. 2, The results.* Cambridge: Cambridge University Press.

Pinker, Steven and Alan Prince
　1988　　On language and connectionism: Analysis of a parallel distributed processing model of language acquisition. *Cognition* 28: 73–193.

Pinker, Steven and Michael Ullman
　2002　　The past and future of the past tense. *Trends in Cognitive Sciences* 6: 456–463.

Rizzi, Luigi
　1994　　Some notes on linguistic theory and language development: the case of root infinitives. *Language Acquisition* 3: 371–393.

Robison, Richard E.
　1990　　The primacy of aspect: Aspectual marking in English interlanguage. *Studies in Second Language Acquisition* 12: 315–330.

Rocca, Sonia
　2002　　Lexical aspect in child second language acquisition of temporal morphology: A bidirectional study. In *The L2 Acquisition of Tense-Aspect Morphology*, M. Rafael Salaberry and Yasuhiro Shirai (eds.). Amsterdam/Philadelphia: John Benjamins.
　2007　　*Child Second Language Acquisition: A Bi-Directional Study of English and Italian Tense-Aspect Morphology.* Amsterdam/Philadelphia: John Benjamins.

Rumelhart, David E. and James L. McClelland
　1986　　On learning the past tenses of English verbs. In *Parallel Distributed Processing: Explorations in the Microstructures of Cognition, Vol. 2: Psychological and Biological Models*, James L. McClelland, David E. Rumelhart and the PDP Research Group (eds.). Cambridge, MA: MIT Press.

Salaberry, M. Rafael
　1999　　The development of past tense verbal morphology in classroom L2 Spanish. *Applied Linguistics* 20: 151–178.
　2000　　*The Development of Past Tense Morphology in L2 Spanish.* Amsterdam/Philadelphia: John Benjamins.
　2008　　*Marking Past Tense in Second Language Acquisition: A Theoretical Model.* London: Continuum Press.

Salaberry, M. Rafael and Dalia Ayoun
　2005　　The development of L2 tense-aspect in the Romance languages. In *Tense and Aspect in Romance Languages*, Dalia Ayoun and M. Rafael Salaberry (eds.), 1–33. San Diego, CA: Academic Press.

Salaberry, M. Rafael and Yasuhiro Shirai (eds.)
　2002　　*The L2 Acquisition of Tense-aspect Morphology.* Amsterdam/Philadelphia: John Benjamins.

Schumann, John H.
　1987　　The expression of temporality in basilang speech. *Studies in Second Language Acquisition* 9: 21–24.

Shirai, Yasuhiro
- 1991 Primacy of aspect in language acquisition: Simplified input and prototype. Ph.D. dissertation, Applied Linguistics, University of California, Los Angeles.
- 1992 Conditions on transfer: A connectionist approach. *Issues in Applied Linguistics* 3: 91–120.
- 1993 Inherent aspect and the acquisition of tense/aspect morphology in Japanese. In *Argument structure: Its Syntax and Acquisition*, Heizo Nakajima and Yukio Otsu (eds.). Tokyo: Kaitakusha.
- 1994 On the overgeneralization of progressive marking on stative verbs: Bioprogram or input? *First Language* 14: 67–82.
- 1997 The L2 acquisition of Spanish and the Aspect Hypothesis. Paper presented at the *17th Annual Second Language Research Forum*, Michigan State University, Oct. 17.
- 1998 The emergence of tense-aspect morphology in Japanese: Universal predisposition? *First Language* 18: 281–309.
- in press Semantic bias and morphological regularity in the acquisition of tense-aspect morphology: What is the relation? *Linguistics.*

Shirai, Yasuhiro and Roger W. Andersen
- 1995 The acquisition of tense-aspect morphology: A prototype account. *Language* 71: 743–762.

Shirai, Yasuhiro and Atsuko Kurono
- 1998 The acquisition of tense-aspect marking in Japanese as a second language. *Language Learning* 48: 245–279.

Shirai, Yasuhiro and Susanne Miyata
- 2006 Does past tense marking indicate the acquisition of the concept of temporal displacement in children's cognitive development? *First Language* 26: 45–66.

Shirai, Yasuhiro, Dan I. Slobin and Richard M. Weist
- 1998 Introduction: The acquisition of tense/aspect morphology. *First Language* 18: 245–253.

Slabakova, Roumyana
- 2001 *Telicity in the Second Language.* Amsterdam/Philadelphia: John Benjamins.

Slabakova, Roumyana and Silvina Montrul
- 2002 On viewpoint aspect interpretation and its L2 acquisition: A UG perspective. In *The L2 Acquisition of Tense-Aspect Morphology*, M. Rafael Salaberry and Yasuhiro Shirai (eds.). Amsterdam/Philadelphia: John Benjamins.

Slobin, Dan I.
- 1985 Crosslinguistic evidence for the Language-Making Capacity. In *The Crosslinguistic Study of Language Acquisition, Vol. 2: Theoretical Issues*, Dan I. Slobin (ed.). Hillsdale, NJ: Lawrence Erlbaum.

1997 The origins of grammaticizable notions: Beyond the individual mind. In *The Crosslinguistic Study of Language Acquisition, Vol. 5: Expanding the Contexts*, Dan I. Slobin (ed.). Mahwah, NJ: Lawrence Erlbaum.

2001 Form-function relations: How do children find out what they are? In *Language Acquisition and Conceptual Development*, Melissa Bowerman and Stephen C. Levinson (eds.). Cambridge, UK: Cambridge University Press.

Smith, Carlota S.
1980 The acquisition of time talk: Relations between child and adult grammars. *Journal of Child Language* 7: 263–278.

Smith, Carlota S. and Mary S. Erbaugh
2005 Temporal interpretation in Mandarin Chinese. *Linguistics* 43(4): 217–256.

Smith, Carlota S. and Richard M. Weist
1987 On the temporal contour of child language: A reply to Rispoli and Bloom. *Journal of Child Language* 14: 387–392.

Stephany, Ursula
1981 Verbal grammar in Modern Greek early child language. In *Child language: An International Perspective*, Philip S. Dale and David Ingram (eds.). Baltimore, MD: University Park Press.

Stokes, Stephanie F. and Paul Fletcher
2003 Aspectual forms in Cantonese children with specific language impairment. *Linguistics* 41: 381–406.

Stoll, Sabine
1998 The role of Aktionsart in the acquisition of Russian aspect. *First Language* 18: 351–377.

Sugaya, Natsue and Yasuhiro Shirai
2007 The acquisition of progressive and resultative meanings of the imperfective aspect marker by L2 Learners of Japanese: Universals, transfer, or multiple factors? *Studies in Second Language Acquisition* 29: 1–38.

Swift, Mary D.
2004 *Time in Child Inuktitut: A Developmental Study of an Eskimo-Aleut Language*. Berlin/New York: Mouton de Gruyter.

Valian, Virginia
2006 Young children's understanding of present and past tense. *Language Learning and Development* 2: 251–276.

VanPatten, Bill
2002 Processing instruction: An update. *Language Learning* 52: 755–803.

von Stutterheim, Christiane and Wolfgang Klein
1987 A concept-oriented approach to second language studies. In *First and Second Language Acquisition Processes*, Carol W. Pfaff (eds.). Cambridge, MA: Newbury House.

Wagner, Laura
    2001   Aspectual influences on early tense interpretation. *Journal of Child Language* 28: 661–681.

Weist, Richard M.
    1986   Tense and aspect. In *Language Acquisition*, 2nd edition, Paul Fletcher and Michael Garman (eds.). Cambridge, UK: Cambridge University Press.
    1989   Time concepts in language and thought: Filling the Piagetian void from two to five years. In *Time and Human Cognition: A Life-span Perspective*, Iris Levin and Dan Zakay (eds.). Amsterdam: Elsevier.
    2002   The first language acquisition of tense and aspect: A review. In *The L2 Acquisition of Tense-aspect Morphology*, M. Rafael Salaberry and Yasuhiro Shirai (eds.), Amsterdam/Philadelphia: John Benjamins.

Weist, Richard M., Hanna Wysocka, Katarzyna Witkowska-Stadnik, Ewa Buczowska and Emilia Konieczna
    1984   The defective tense hypothesis: On the emergence of tense and aspect in child Polish. *Journal of Child Language* 11: 347–374.

Wexler, Kenneth
    1994   Optional Infinitives, head movement, and the economy of derivation in child language. In *Verb Movement*, David Lightfoot and Norbert Hornstein (eds.). Cambridge: Cambridge University Press.
    1998   Very early parameter setting and the unique checking constraint: A new explanation of the optional infinitive stage. *Lingua* 106: 23–79.

Wexler, Kenneth and Peter Culicover
    1980   *Formal principles of language acquisition*. Cambridge, MA: MIT Press.

Wu, Ruey-Jiuan Regina
    2002   Discourse-pragmatic principles for temporal reference in Mandarin Chinese conversation. *Studies in Language* 26: 513–541.

Yap, Foong Ha, Stella Wing-man Kwan, Patrick Chun-kau Chu, Emily Szeman Yiu, Stella Fat Wong, Stephen Matthews and Yasuhiro Shirai
    in press   Aspectual asymmetries in the mental representation of events: Role of lexical and grammatical aspect. *Memory & Cognition*.

Yoshitomi, Asako
    2007   Testing the Primacy of Aspect and Reverse Order Hypothesis in Japanese returnees: Toward constructing a corpus of second language attrition data. In *Corpus-Based Perspectives in Linguistics*, Yuji Kawaguchi, Toshihiro Takagaki, Nobuo Tomimori and Yoichiro Tsuruga (eds.). Amsterdam: John Benjamins.

New perspectives in analyzing aspectual distinctions across languages

Christiane von Stutterheim, Mary Carroll and Wolfgang Klein

> Some physicists believe we should think the unthinkable and abolish time altogether.
> Michael Brooks (2008: 32)

1. Introduction

There is a clear distinction between an **event** – the word is taken here in a broad sense[1] – and the **description** of a given event by some speaker on some occasion. A famous murder in the antique world involved two participants, Brutus and Caesar, some activity on the part of the former and some changes that happened to the latter: first, he was alive, and then, he was not alive. These and other features of the event – the time and the place at which it happened and the instrument which Brutus used – are part of reality. A speaker, faced with the task of giving a fairly accurate description of this event, has a number of options. He may or may not choose to take into account some of the entities – persons and objects – that are involved:

(1) a. Brutus killed Caesar with a dagger.
    b. Brutus killed Caesar.
    c. Caesar died.

He may decide to present this event as on-going, or as completed:

(2) a. Brutus was killing Caesar with a dagger.
    b. Brutus killed Caesar.

In this case, the specific language, English, provides the speaker with two verbal forms (a "continuous form" and a "simple form") which differentiate between these two options. Other languages use adverbials or periphrastic constructions of a different type to this end. The speaker may present the description in one clause, or distribute it over several clauses or sentences:

---

[1] It includes all sorts of situations that can occur or obtain in the real or in a fictitious world, including states, processes, actions, or events in a narrower sense, etc.

(3) a. Brutus killed Caesar.
b. Brutus decided to kill Caesar for political reasons. He carried out his plan immediately. He took a dagger and stabbed Caesar several times. Caesar died on the stairs of the Curia.

There are many other options in giving a description of one and the same event. The speaker can add all sorts of additional features, from the facial expression of the protagonist to the shape of the instrument involved. But these options are not unrestricted. There are at least four types of more or less strong constraints that influence or determine the speaker's options:

A. A first type of constraint comes with the particular communicative task that the speaker wants to perform. The nature of this task determines the choice of features that are made explicit and the level of detail – the "granularity" – with which the facts are presented (see von Stutterheim 1997).
B. A second type of constraint is given with the lexical and grammatical properties of the particular language used. Some languages, English, for example, force their speakers to indicate the time at which the event occurred, since tense marking on the verb is obligatory. The speaker must therefore locate the event in the past, present or future. Other languages, such as Chinese, leave it to the speaker's discretion to provide information about the "when" of the event. Similarly, some languages have a form that is neutral with respect to "on-going" and "completed"; other languages force their speakers to choose between one of these options, while another set force them to make an aspectual choice for one tense but not for another.
C. A third type of constraint is given with the individual level of proficiency with which the means provided by the language are used: there are highly proficient speakers, and ones who are less so. It is one thing to know what is structurally and lexically possible, and quite another matter to use this potential in solving a particular communicative task. It is not enough to have a tool box, one must also be able to use it. This is particularly clear in the case of language learners who are still on their way to becoming more or fully proficient.
D. Finally, speakers may also be constrained in the choices they make by particular cognitive, cultural or social habits. Although someone could present a certain event in a certain way, he will be unlikely to do so because this may not correspond to the way he, or the community to which he belongs, would normally relate to it. This concerns, for example, the

choice of information which is considered relevant for a particular task, or what should be made explicit or left implicit in the message. Typically, these constraints are not of an obligatory nature; they simply propel the speaker in a certain direction when formulating a message, while still leaving a number of options.

One such option is the choice between presenting a given event as "ongoing" or as "completed", illustrated above for English in (2a) and (2b), respectively. Traditionally, these two ways of presenting something are called "aspects", with the basic distinction between "imperfective aspect" and "perfective aspect". There is a vast and steadily increasing amount of research on the general notion of aspect and on these two aspects in particular (see Klein, this volume, section 3). Very roughly, there are three main lines of study that are interconnected in many ways. There are descriptive accounts of a number of aspectual systems, often treated in the context of the entire verbal system. If aspect is grammaticalized in a language, then any descriptive grammar of that language will say something about its form and meaning – as, for example, the many in-depth studies of the English progressive. Second, there are crosslinguistic and typological studies in which the expression of aspect is compared across a more or less extensive range of languages; the comparisons may be confined to a particular language, such as Slavic languages, where the differences at issue are usually quite subtle. There are also comparisons of languages that are typologically very different in which forms as well as meanings of aspectual markers diverge considerably, thus giving rise to the question of finding a common denominator for this category is (see, for example, Dahl 1985, 2000; Smith 1997). Third, there is a host of studies, many of them in formal semantics (see von Stechow, this volume), which address theoretical issues on aspect, or particular aspectual forms; one cannot claim that there is a shortage of theories on this topic (see, for example, Smith 1997 or Rothstein 2008).

In view of all of this work it is surprising how little agreement has been reached as soon as one goes beyond the most elementary level of description. This applies both to accounts of the empirical facts as well as the underlying theoretical notions. It holds even for languages that have been intensively studied in this regard, such as English or Russian. In both languages, aspectual distinctions are grammaticalized, and the morpho-syntactic facts of the distinction are well described. In both languages, the differences in meaning between aspectually marked forms involve, in a broad sense, the familiar and well-established notions of imperfectivity and perfectivity:

A. The situation is presented "from outside" versus the situation is presented "from inside".
B. The situation is presented as "completed" versus the situation is presented as "non-completed" or "on-going".
C. The situation is presented "with its boundaries" versus the situation is presented "without its boundaries". (Klein, this volume, section 3)

But clearly, the Russian imperfective and the English progressive are not equivalent. Any translation of a Russian novel into English, or an English novel into Russian, makes this unmistakably clear; and any English speaker who has ever attempted to learn Russian, or any Russian speaker who has tried to learn English, is confronted with differences not only in meaning but also in principles of use. If this is true for the imperfective in Russian and English, to what extent can we learn from descriptions of imperfective aspect in Chinese, for example, for which there are nevertheless quite a few empirical studies (see, for example, Xiao and McEnery 2004), compared to the many languages for which we have no more than one or two grammars written for practical purposes by a missionary or an isolated fieldworker? Aspectual characterizations may not be incorrect – but they are definitely not fine-grained enough.

We believe that there are two main reasons for the present state of research in the field of aspect (and probably in the field of temporal expressions in general). They are both related to the way in which aspect is typically investigated. First, these investigations focus on the meaning of particular morpho-syntactic forms, for example the attachment of the (perfective) prefix *po-* in Russian or the *-ing* in English, and the combination of the resulting form with a copula, for example. With the focus on meaning and form, use of these forms is not analyzed in the context of the more complex processes given with the verbal description of specific events. This process is determined by a number of factors, in particular those mentioned above under A–D, as well as the interplay of these and perhaps other factors in determining the way in which a speaker "presents an event". Secondly,- and this point is related to the first – the traditional ways in which the relevant data are defined, as well as the way which the necessary empirical evidence is collected, restrict the empirical facts taken as the starting point for theoretical deliberations (see section 2 below).

In this paper we present a method of analysis whereby established research procedures in the field of time semantics are complemented by experimental tasks. Speakers of different languages are asked to respond to an identical task (view a structured set of situations), with an input that is visual and non-verbal, and thereby give a verbal account of what they have seen. This elicitation technique allows for the manipulation and analysis of the different steps involved between the perception of a particular visual stimulus on the one end and the utterance of a speech signal on the other. In the studies in question, speakers with different language backgrounds are asked to describe the same set of events. This event can be simple – a scene presented in a short video clip; but it can also be complex, such as the account of a car accident, which involves many sub-events that are temporally related to each other, with changes in places, objects, persons and the properties which these objects and persons have. Speakers are thus confronted with a verbal task in which all or many of the options outlined in (1)–(3), but also all or many of the constraints mentioned under A–D, come into play and can be investigated. It is also possible to study how speakers solve this task in real time, for example, by taking into consideration chronometrical parameters such as speech onset, or eye movements while viewing the scene, or in the course of language production, or both. Factors of this kind reflect the way in which attention is directed to the different components of the event. It is also possible to systematically analyze the final product – the sentence or sequence of sentences in which the event is described. This allows the investigation as to how patterns in event construal are adapted to the particular tools which a particular language offers. In other words, if there is a form of "language specificity", that is, a systematic influence given by the linguistic system in event representation, then this should become evident in the way in which speakers deal with one and the same task.

In illustrating the empirical approach we will focus on one aspectual distinction – the speaker's option of presenting an event as *on-going* or as not *on-going*. As mentioned above, this distinction is preliminary at this stage, and does not reflect the variety of actual aspectual meanings in different languages. We will report on findings for the Germanic languages English, German, and Dutch, as illustrations of the method used in investigating aspect. The methods can also be applied to the analysis of different temporal expressions, such as tense or temporal adverbials, and in particular to the interplay of these devices.

2. Some background considerations

As in any scientific endeavor, the study of how temporality is expressed is embedded in a given research tradition which shows an accumulation of an impressive body of knowledge. Any tradition may prove to be an obstacle to a better understanding of the domain if it precludes the adoption of a fresh and unbiased look at the phenomena under investigation: we benefit, but we also suffer from tradition. It is thus useful to reflect briefly on the empirical and theoretical underpinnings of this tradition before taking up the thread once again – perhaps with a slightly different view.

Let us begin with the empirical side. What is the nature of the evidence on which the linguist's claims about the lexical and grammatical features of a human language are normally based? Essentially, there are two methods that are generally not strictly isolated but combined in different forms:

- looking at examples of language use, that is, written or spoken sentences or texts in a given language; this may be carried out in a systematic way by compiling and analyzing a data corpus; it may also be carried out on an ad hoc basis.
- appealing to the linguistic "intuitions" of someone who speaks the language (very often the linguist's own intuitions). This appeal can take the form of a grammaticality judgment (e.g., "Can you say that in Sorbian?") or a question about meaning (e.g., "What do you use this word for?").

Take, for example, a question such as: "What is the position of the finite verb in German, compared to Latin or Turkish?" A common way of answering this is to look at a more or less representative corpus of sample sentences. An initial inspection shows that the position of the verb in Latin is relatively free – it can appear at the beginning of the clause, at the end, and somewhere in-between. In Turkish, it is predominantly at the end – although other positions are found as well. In German, it also appears in various positions: *Hans stelle die Tasse auf den Tisch. – Stell die Tasse auf den Tisch! – ..., weil Hans die Tasse auf den Tisch stellte.* Other possible positions are not observed. Both findings – the positions that are observed in the data, and the positions that are not observed in the data – are not fully satisfactory in answering the question. It could be accidental that certain positions are not observed (especially if the corpus of sample sentences is small). This can in part be remedied by changing the position and asking someone who speaks the language whether the resulting structure is possible – an appeal to the "intuitions" of this speaker. And as to the positions that are in fact observed – they yield a somewhat inconsistent picture. It is easy to see that the various positions are not random, but correlate somehow with structural and functional differences. But what are these differences? Here, the linguist resorts to his or her intuitions as someone who speaks the language. In this particular example, it is possible to correlate varying positions with varying types of speech situations in which the utterance is used, and then relate the different positions to different functions.

An observational approach of this kind is hardly possible when it comes to the meaning of words, the meaning of bound morphemes such as a suffix, or the meaning of full constructions. In theory, one might try to determine the meaning of *to kill* or *to die* by observing all types of situations in which these words are used by someone (and in fact, this is essentially the way in which the meaning of these expressions is learned). But first, this is practically impossible if more than a few words are to be described. Secondly, it is not very revealing, because one cannot easily determine which feature of the situation is related to that particular word. Finally, it is plainly impossible with expressions with a more functional meaning, be it morphemes such as *-ed* or *-ing*, or full words such as *the, still* or *or*. In practice, and sometimes in principle, there is so far no other option than to appeal to the intuitions of a competent speaker of the language in question; this appeal is supported by the clever use of specimens of the language when used in production. These can take the form of combinatorial tests. One might ask speakers (or oneself), for example, whether it is possible to combine a particular verb with the adverbial *for two hours*, and when these speakers' semantic intuition tell that this is not possible for *to explode* but possible for *to sleep*, we might take this fact as a basis for verb classification. This may be helpful, but only within certain limits. It cannot be readily transferred to other languages. And even in English, we have no satisfactory answer as to why such a combinatorial restriction applies and whether it is important (see Klein, this volume, section 4).

All of this is well-known. But if taken seriously, then it should not be surprising that our knowledge of how lexical aspect, grammatical aspect, as well as other temporal devices function in particular languages is still incomplete. Is there any way, any instrument, any measurement or procedure that would allow us to go substantially beyond what these two methods can provide us with? In fact, when it comes to meaning it is difficult to imagine how one could proceed without an appeal to the semantic intuitions of someone who "knows" the language. The meanings of *to work, beyond* or *-ing* are not visible, neither on paper nor in sound waves nor in the brain. But we can **refine** these two classical methods, in order to introduce a better level of control and comparability, and most importantly, to look at what speakers of some language actually do when describing a scene and use certain words and constructions to this end. This is the aim of the methods discussed in the following sections.

Let us now turn to the theoretical side. There are numerous, often very sophisticated and formally very elaborate theories of aspect. However, we are very far from a generally accepted theory of, say, the perfective-imperfective distinction. The aim of the following section is not to present another theory, but to point out some assumptions which theories have to consider. We will keep these assumptions to the minimum.

Consider 2 again, repeated here:

(2) a. Brutus was killing Caesar with a dagger.
    b. Brutus killed Caesar.

These are two possible descriptions of one and the same situation. Intuitively, they have certain elements in common, and they differ in other respects. In both cases,

– there must be a person who does something, for example, uses a dagger; this person – the "agent", encoded here as subject – is called Brutus;
– there must be someone who is first alive and then dead; this person – the "patient", encoded here as the direct object – is called Caesar;
– the change of properties in the case of Caesar is caused by Brutus' activity;
– there is a inherent temporal structure; the time at which Caesar is alive precedes the time at which Caesar is dead, and the time at which Brutus does something must overlap with the time at which Caesar is alive; so, we have at least three inherent temporal intervals which are characterized in a particular way.

Note that the same inherent temporal structure is also found in sentences such as *Peter opened the door* or *Mary was painting the wall*. What differs are the entities that fill the argument positions, and the properties which go with the various intervals: doing something with the door handle, or the brush, rather than with a dagger; being first closed and then open, not green and then green; being alive and then dead. Thus, *to kill, to open, to paint* have the same argument structure, and they have the same inherent temporal structure, but the descriptive properties assigned to the arguments and the various temporal intervals which make up this structure are different. Other verbs, such as *to die, to cost, to watch, to give* may also have a different argument structure, and they may have a different temporal structure, and, of course, different property assignments to the various arguments at the various time spans.

These considerations, simple as they are, lead to three important conclusions that must be observed in the study of temporal expressions:

1. Even with a simple verb, there can be an inherent complex temporal structure; in the examples just discussed, there are (at least) three temporal intervals which are related to each other by relations such as *later* or *overlapping*.
2. There is a clear difference between the *inherent temporal structure* given with an expression and the *descriptive properties* that go with the components of this structure, i.e., the various temporal intervals.
3. Temporal structure and descriptive properties of a lexical verb (and a more complex construction based on a verb) are relative to the arguments. The event encompasses different subintervals with different properties of the entities involved: it is Caesar who is dead at some subinterval of the event, whereas he was alive at some earlier subinterval of the same event.

Since all verbs do not require two arguments, they need not involve two time spans. In *Caesar died*, we have only one argument, but two time spans with the same descriptive properties, as specified for the second argument in *Brutus killed Caesar*: at the first time, he was alive, and at the second time, he was dead. The verbs *to kill* and *to die* are both what one might call "two state verbs"; but note that these two states – the source state and the target state or end state – are relative to specific arguments, here the second argument in *to kill* and the only argument in *to die*. Between these two states, there must be a transition, a point of transition, so to speak. The verb itself does not say anything about whether this transition is abrupt or smooth; so, the "transition point" need not really be point-like. In other verbs, the distinction between source state and end state is only gradual, as, for example, in *to rise*; if the temperature rises, it need not be "low" in the source state and "high" in the end state – it should only be "higher" in the second state than in the first. Note that this difference does not concern the temporal structure – in *to die* as well as in *to rise*, it involves a first time and a second time – but the descriptive properties which go with it: the difference between them can be categorical (be alive – be dead), it can also be gradual (lower – higher). The precise analysis of how this works with specific verbs is complicated and goes beyond the subject of this paper. But any analysis must somehow deal with these elementary observations.

Let us now return to the difference between *Brutus killed Caesar* and *Brutus was killing Caesar*. What has been said so far applies to both of these two (out of the many) options of describing one and the same event. Intuitively, the second version – the progressive – places us somehow in the midst of the action and the action is presented as "on-going". This does not preclude that it comes to an end; but the sentence *Brutus was killing Caesar* could be true without Caesar actually reaching the stage at which he is dead – the "end state" or "target state" of the entire action; someone could have intervened in due time. In *Brutus killed Caesar* this is not possible. The end-state or target state must indeed be reached, or the speaker's description is wrong. We can easily catch this intuition if we assume that the speaker's assertion is confined to a particular time – the time about which the speaker wants to talk, in this case, the time for which the assertion holds. In *Brutus was killing Caesar* this time must overlap with the time of Brutus' activity, but not the time of at which Caesar is dead – the end state of the second argument. In *Brutus killed Caesar* the time at issue must overlap with the time of Brutus' activity as well as the two other time spans – the time at which Caesar was alive (you cannot kill a dead man) and the time at which he was dead. In both cases the verb meaning of *to kill* includes an end state – a temporal of "Caesar be dead"; but in saying *Brutus was killing Caesar*, the speaker is not committed to the claim that this end state is reached; in *Brutus killed Caesar* he is committed to such a claim.[2]

Again, these observations are not particularly sophisticated. But they show that if we want to understand how aspectual distinctions function, we should not only look at the inherent temporal structure of verbs (and more complex verbal expressions), but also at the time about which the speaker makes the statement. Following Klein (1994) we shall call this time the "topic time". In the special case of assertions, we could also call it the "assertion time", i.e., the time to which the speaker wants to confine his statement. This choice is one of the many options which a speaker has when he or she sets out to describe a particular situation in a particular communicative context. It interacts with the lexical information provided with the verb – its inherent "argument-time structure" and the descriptive properties that go with it. It also interacts with the contribution made by other expressions to the entire meaning of a sentence, a point which we shall not take up here.

These background considerations serve to point to some of the facts which one has to keep in mind when investigating aspectual distinctions, as

---

[2] Note that the sentence need not be true; the speaker could be wrong, or telling a lie.

mentioned earlier. In the next three sections we will illustrate how this can be carried out using empirical methods of analysis. We will show how speakers of three closely related languages, English, Dutch and German, proceed when describing motion events, as presented in video clips. We will then take a more differentiated look at other situation types and the extent to which they attract use of an aspectual viewpoint in these three languages.

3. Looking at the endpoint: the case of English, Dutch and German

Among the differences between the three main West Germanic languages, the degree to which aspect is grammaticalized is one of the most salient. In English there is a systematic distinction between the meaning conveyed by a progressive or continuous form, as in *John was closing the window, Peter was sleeping*, and forms which encode the completion of the event as in *John closed the window* or *John has closed the window*. With a few exceptions, this distinction applies to all verbs, and the speaker must choose between them when presenting an event: there is no neutral form (a good empirical survey is given in Williams 2002). German, on the other end, has no grammatical distinction of this kind. The simple forms *Hans schloß das Fenster, Peter schlief* can have both readings, and if he speakers wants to mark "on-goingness", then he or she has to use adverbials (such as *gerade* "just") or periphrastic constructions. The most important of these involve the prepositions *an* or *bei*, which both mean (roughly) "at". But the use of these forms is infrequent[3] and limited to some types of verbs: *Hans war dabei das Fenster zu schließen*, lit. "John was at-it the window to close" is fine, whereas *Peter war dabei zu schlafen, Peter war am Schlafen* or *Peter war beim Schlafen* are interpretable, but odd. Dutch is somewhere between these poles. Aspect is not grammaticalized, as in German, in the sense that its use is not obligatory in given contexts. The most common form to express the on-goingness of an event is the construction *aan het* + infinitive (*at the* + infinitive), which occurs with *zijn* (to be) as in *een man is viool aan het spelen* 'a man is violin <u>at the</u> play'. However, use of the associated aspectual perspective is considered to be much more common than in German (van Pottelberge 2004 is a comprehensive survey of the facts; see also Boogaart 1991; Booij 2002).

---

[3] It is, however, considered typical for some German variants close to the Dutch border ("niederrheinische Verlaufsform", i.e., lower-Rhine progressive form).

This is no more than a rough sketch of the facts. It could be made more precise by (a) asking native speakers for their intuitions – both about what on-goingness means, whether they would use a particular form with a particular verb, or if they would rather use this or that form in a given situation, or (b) by looking at the use of these forms in actual corpora or spoken or written language. Both methods have their merits and their shortcomings (cf. section 2). In what follows, we will adopt a different approach – we will look at how English, Dutch and German speakers proceed when asked to describe one and the same set of events presented as on-going in a video-clip. The first set presented here involve situations showing *motion events* with a change in place such as *a person is walking down a street* where the end-state is a place (for example *a person walking to a house*). In traditional terms one could say that an event of this kind has a "right boundary". But the following points must be noted here. First, the "boundary" is not just the end of an event since it is characterized by certain properties assigned to one of the arguments. These properties can be spatial (the entity is AT the house, whereas it was not before) – it is an end-place. The property could be of a more qualitative type (*the wall is painted*); a special case of the latter kind is that the referent of the argument only exists at the "target state": the house is now built, whereas it did not exist before. Furthermore, a difference must be made between whether the end-state is **mentioned** and thereby **marked as reached**. In English, for example, the description of one and the same scene could be:

(5)   a. Two nuns were walking.
        b. Two nuns were walking to a house.
        c. Two nuns walked to a house.

In (5b) as well as in (5c), the end-state – in this case the target position of the only argument, the nuns – is made explicit; but only in (5c), the time about which an assertion is made (the topic time) is large enough so as to include the end-state. If the end-state is not shown in the video as reached, the speaker need not necessarily zoom in on exactly what is shown – he or she could construe this phase as a part of a larger event, which includes the end-state. Then, the speaker has, so to speak, a holistic representation of the event: it is "viewed" in its entirety, although it is not presented in the video in its entirety.

English, Dutch and German speakers (20 per group) were shown 80 video-clips and were asked to tell *what is happening?* They were also asked to start as soon as they recognized what was going on. 18 of the clips (test items) showed situations involving a change in place of a figure (person or vehicle), while the rest were so called distracters. The speaker's utterances were tape-recorded and categorized by two independent observers as to whether the description included a reference to an "endpoint" of whatever type; this could include all sorts of end-states, spatial, qualitative, existence.[4]

3.1. Referring to the endpoint

The results for the three groups of speakers show a clear difference with respect to the verbalization of endpoints.

Table 1. References to endpoints of motion events

English	Dutch	German
23.3%	15.5%	76.4%

English speakers tend to focus on the phase which is actually presented in the clip. If endpoints are not reached by the figure in motion, there is a clear preference not to mention them – although this is clearly possible, as in (5b) above. The –ing form is used in all cases (100%). In other words, since they were asked to tell what is happening, the event shown in the clips is perceived as on-going and the "zooming-in" on this phase goes so far that they do not mention a possible endpoint.

The results for German give a different picture. First of all, the event is not presented as on-going, although the system provides means to express this perspective. But constructions such as *am V-inf sein* or *beim V-inf sein* are not used to describe motion events, nor do they use adverbials to this end. They do not relate to on-goingness as a feature of events. Second, they show a clear tendency to refer to the endpoint, whether it is presented in the clip or just inferable. They do not "zoom into" a phase of the event but construe it as a whole; this holds even when it is left open as to whether the endpoint is actually reached within the time interval of the video clip. In

---

[4] We have chosen to use the term "endpoint" in referring to the final point of a motion event since the term "(right) boundary" is used in different ways in the literature (see Klein, this volume). The term "end-state", on the other hand, is normally not associated with the final destination of a motion event.

other words, they take a holistic view on the events in contrast to the strategy of phasal decomposition in English.

Dutch differs from German in that on-goingness is indeed coded by *aan het V-inf zijn*. But in contrast to English, use of this perspective on-goingness is relatively low (see section 4 below). Restrictions in the use of on-goingness in Dutch can be explained by the fact that the relevant morpho-syntactic construction is still evolving in the system. The grammaticalization of the periphrastic form *aan het* + infinitive in Dutch is reflected in the syntactic status of the prepositional phrase: in *een man is viool aan het spelen* the preposition *aan* is adjacent to the nominalized infinitive *het spelen*; the direct object *viool* precedes the entire prepositional phrase. This suggests that that the preposition is no longer head of a phrase which includes the object (see also Boogaart 1991; Booij 2002); it is "getting closer" to the verb, thus weakening its status as a locative preposition and approaching the status of an independent grammatical construction for on-goingness. Its semantic interpretation as a locative phrase governing a nominal phrase is bleaching into a verbal element; in this sense, it is becoming grammaticalized. The two constructions *viool aan het spelen zijn* and *viool spelen* are already quite parallel (note that in Dutch, as in German, the direct object precedes the non-finite verb). In German, by contrast, the prepositions have retained their locative status: *er ist dabei Violine zu spielen* "he is there-at violin to play" or *er ist beim Singen* "he is at-the singing"; one cannot say *er ist Violine am Spielen* or *er ist ein Lied beim Singen*.

Summing up, English, German and Dutch speakers show different preferences in the way the same events are conceptualized from a temporal perspective, in particular in the way they relate to the endpoint. The findings were investigated further with experiments testing non-verbal behavior: chronometric analysis of speech onset times, as well as eye tracking while viewing the clips.

3.2. Comparison of speech onset times

This study is based on the same data. Speakers had been asked to start speaking as soon as they recognize what is going on in the video clip. The hypothesis was that if German speakers need an endpoint in order to satisfy the notion as to what can be coded as an event, they will wait for the event to become evident as a whole before starting to speak. By contrast, speakers of English, in which the aspectual distinction of on-goingness is coded grammatically, can describe any phase of an event (inceptive, intermediate, terminative) and therefore do not have to wait for the endpoint an action.

Speech onset times for the test items were determined by measuring the onset of the sound wave in the digitalized version of these recordings.

Table 2. Speech onset times

English	Dutch	German
3.51 sec	4.0 sec	4.54 sec

On average, German speakers started speaking 4.54 seconds after stimulus onset (i.e. after the beginning of the video clip), while English speakers started about one second earlier ($t1(24) = 3.13$, $p = .004$; $t2(27) = 10.71$, $p < .001$). Dutch speakers are in between. This means that in order to get a reportable event, German speakers indeed wait longer, while English speakers are able to verbalize an on-going event such as *someone is walking* without having to figure out the endpoint (where the person is heading, for example). The results thus confirm what was found in the analyses of the verbal tasks: in providing the basis for a reportable event, German speakers show a clear preference for a holistic perspective, and this means waiting until the scene as a whole has unfolded before starting to speak. In contrast, any phase constitutes a reportable event for English speakers.

3.3. Eye tracking

The specific hypothesis in this case is as follows: When verbalizing information on situations showing a figure on its way from one place to another, but where the goal in the video clip is not actually reached during the span observed, speakers of languages that do not have grammaticalized means to represent an event as on-going will not only refer to endpoints but will also be more likely to attend to them, compared to speakers of languages in which the temporal-aspectual concept "event is on-going" is grammaticalized.

Table 3. Number of fixations of endpoints before and after speech onset

	English	Dutch	German
Fixations before SO	2.9	4.06	6.9
Fixations after SO	8.5	5.59	9.5

The figures show a marked difference between German and English speakers in the number of fixations before speech onset, while Dutch is in between (see also von Stutterheim et al. 2008, with results for seven languages). The German speakers require more information on the event before beginning to speak, since the event is construed in holistic terms with a possible endpoint.

Why is there such a pronounced difference between fixations before and after speech onset for the English group? English speakers apparently start speaking before they focus the endpoint and, as mentioned above, this can be linked to the fact that if motion events are viewed in aspectual terms, they are typically decomposed into phases (inceptive, intermediate, terminative phase). As with speech onset, the underlying phasal structure explains why English speakers – in contrast to speakers of German – do not tend to scan the scene for an endpoint from the outset. However, the eye tracking results show that in the course of the scanning process they visually control for possible endpoints. Speakers can add on the terminative phase and easily integrate it into the sentence which is already underway: *A car is going down a lane...... to a farmhouse*. These results provide evidence of a language-specific effect at the level of conceptualization, that is, before the speakers begin to speak.

4. Presenting an event as on-going: Dutch and German

We have seen how speakers of English use an aspectual viewpoint when asked to tell "what is happening" and thereby respond to the phase focused in the video clip (intermediate phase). This means that they defocus the terminative phase, at first at least, and do not typically relate to the endpoint in the information verbalized. The questions pursued in the present section of the study are as follows:

– Are speakers of Dutch and German more likely to use an aspectual viewpoint in situations that do not have a prominent endpoint or transition point, as in situations listed under A below?
– How do they proceed if the situation has a prominent transition point with a target state, as in type B?

Type A Relatively homogeneous situations which last for a while and do not involve a salient change of any of the entities involved, e.g. a person surfing, jogging, or swimming;

Type B Situations which involve a salient "qualitative" change with respect to an entity (an effected object, as in building a monument, knitting a scarf; making a paper airplane); the scenes all show progression toward a qualitatively characterized target state;

Type C Motion events (going to the station) were again included in order to examine the extent to which they are viewed as on-going in both languages;

Type B has what one might call an "inner transition" with progression to a target state. This is given by the inferable final state of the object coming into existence, so to speak, whereas this is not the case for A. All three types are reflected in familiar Aktionsart classifications found in the literature (see Klein, this volume, section 4). It should be emphasized, however, that the classification into types, as given above, relates to situations shown in the video clips and not to distinctions which may be captured by linguistic expressions. So we are not talking about differences between verb types but situation types presented in video clips. This method allows us to systematically manipulate specific temporal properties of the events depicted so as to find out how speakers of the different languages respond to the same set of stimuli and the specific temporal features presented in them.

In contrast to English, where use of an aspectual perspective is 100% when viewing the same set of situations and telling what is happening, speakers of Dutch and German may represent the event as on-going, or not, as the following overview shows.

Table 4. Use of the aspectual viewpoint 'event is on-going' in Dutch and German

20 speakers	Use of an aspectual viewpoint
German	5.83%
Dutch	30.74%
English	100.00%

Dutch has other locative means, in addition to the periphrastic form, which include the verbs *zitten* (sit) and *staan* (stand), but use is relatively low in the present study (see also van Pottelberge 2004):

Table 5. Different aspectual constructions used – Dutch (% of total no. of responses)

	Aan het + V-inf zijn	Zitten/staan te + V-inf
L1 Dutch	237/911 – 26.02%	43/911 – 4.72%

We will now compare how one factor, the type of situation, in the sense outlined above, influences the choice made when taking a temporal perspective in Dutch and German.

Table 6. Viewing situation types as on-going in Dutch and German – situation types

20 speakers	Dutch	German
Type A (swimming)	(66/151) 43.71%	(20/151) 13.24%
Type B (molding a vase)	(83/190) 43.68%	(8/190) 4.20%
Type C (going to the station)	zero	zero

English is again at 100% for all three situation types.

Type B situations were divided into groups with homogeneous sub-events (*knitting a scarf*), and those with a higher range of heterogeneous subevents (*folding a paper airplane; baking a cake*), in order to test the role of homogeneity in adopting an aspectual viewpoint. In the case of knitting a scarf, the action shown in the video focused on the homogeneous movement of the hands. In the other group (such as baking a cake) different actions were shown such as stirring the cake mix and adding in flour, or folding the paper in different ways, straightening the wings, etc. The results reveal that B type situations with homogeneous sub-events have the highest rate in attracting use of an aspectual perspective.

Table 7. Type B Situations with homogeneous versus heterogeneous sub-events (5 video clips each)

Aspectual perspective	Dutch	German
Type B Homogeneous subevents	48/76 63.16%	3/76 3.94%
Type B Heterogeneous subevents	17/76 22.37%	2/76 2.63%

The results for Dutch indicate that events showing a progression toward a resultant state (type B), and sub-events that can be viewed as homogeneous, are the most likely candidates in leading to the selection of an aspectual perspective.

Finally, the very few motion events (type C) that are represented aspectually as on-going in Dutch are confined to situations that can be viewed in a way similar to type A situations. The situation is represented without a directed change in getting from a specific point of departure to a possible goal. The event is viewed as "being out for a walk", with no mention of any goal, using a verb such as "wandelen". These occurrences were not included in the figures for goal oriented motion events (type C). It should be emphasized at this point that the mention of endpoints in motion events in the Dutch data may fluctuate depending on the extent to which speakers adopt an aspectual perspective in describing the set of scenes, as the entire series of other experiments on motion events have shown (see also von Stutterheim et al. 2008 (submitted).

4.1. Discussion of the results

These findings give rise to a number of considerations. Dutch speakers are most likely to represent an event as going when the scene shows progression toward a qualitatively characterized target state – when a series of causative actions leads to the existence of a finished object through a progressive change in its qualitative properties (type B). Speakers also view the situation as on-going when the event is homogeneous and shows duration, without a progressive change, as in surfing or swimming situations (type A). Homogeneity is a core factor in type B as well as type A events, as the comparison between situations with homogeneous versus heterogeneous events (type B) show.

Although figures are still very low for German, there is emerging evidence that type A situations also attract use of an aspectual perspective, but not when the situation shows a progressive property leading to a resultant state. The findings thus indicate that the aspectual distinction encoded in Dutch encompasses what may be called "progressive" as well as "on-going" components, while the relevant criterion in German may centre on on-goingness – as represented by homogeneous events with a clear duration and no qualitative change.

Finally, the very few motion events (type C) that are represented aspectually as on-going in Dutch are confined to situations that can be viewed in a way similar to type A situations. The situation is represented without a progressive change in getting from a specific point of departure to a possible goal: it is viewed as "being out for a walk" with no specific goal using verbs such as "wandelen". Representation in this way conforms with a possible constraint that calls for homogeneity and duration, but not changes leading to a resultant state. So why is the form of progression (change in place from source to goal) given with motion events not yet accessible for use with an aspectual perspective, in contrast to causative (type B) events? The answer may lie in the nature of the transition. In contrast to motion events, causative actions, as presented in type B situations, may constitute a prototypical context for the concept of progression, since they provide a tangible contrast between the pre-state of the event and its progression to the final post state.

The results presented here for Dutch and German form part of a larger study on aspectual concepts in Germanic (Flecken, submitted), Romance (Natale 2008; Carroll et al. 2008; Leclerq 2008) as well as Semitic (Arabic) and Slavic languages (v. Stutterheim et al., submitted).

5. Concluding remarks

Aspect is an important, but also a very difficult temporal category, and studies on how it should be defined and on how it is realized in different languages are legion. But we are far from reaching agreement on what is involved, except on a very global level, and our knowledge about the form and function of aspects in particular linguistic systems is far from satisfactory. Statements such as "language x is an aspect language" or "language y has an imperfective aspect", may not be false, but they hide more problems than they answer. The aim of this chapter was to present an empirical approach to the study of aspectual phenomena. Rather than depending on the semantic intuitions of someone who knows the language, or examinations of the use of aspectual forms in corpora, empirical analyses were conducted on how speakers proceed when solving different verbal tasks. Tasks can be systematically modified in many ways, and with this the range of analysis of the resulting data. The few findings presented here indicate how a procedure of this kind may lead to a more refined picture of aspectual distinctions, compared to traditional methods. It can not replace these methods, the appeal to intuition and the examination of corpora remains indispensable, but they have their limits and research procedures should try to go beyond them.

References

Boogaart, Ronny
    1991   "Progressive aspect" in Dutch. In *Linguistics in the Netherlands,* Frank Drijkoningen and Ans van Kemenade (eds.), 1–9. Amsterdam: Benjamins.

Booij, Geert
    2002   Constructional Idioms, Morphology, and the Dutch Lexicon. *Journal of Germanic Linguistics* 14: 301–329.

Brooks, Michael
    2008   What makes the universe tick? *New Scientist Magazine* 2683 (22 Nov. 2008): 32–35.

Carroll, Mary, Silvia Natale and Marianne Starren
    2008   Acquisition du marquage du progressif par des apprenants germanophones de l'italien et néerlandophones du français. *AILE*, 31–50.

Dahl, Östen
    1985   *Tense and Aspect Systems*. Oxford: Blackwell.

Dahl, Östen (ed.)
    2000   *Tense and aspect in the languages of Europe.* Berlin/New York: Mouton de Gruyter.

Drijkoningen, Frank and Ans van Kemenade (eds.)
    1991   *Linguistics in the Netherlands.* Amsterdam: Benjamins.

Flecken, Monique
    2008   Event conceptualization by early Dutch-German bilinguals: Insights from linguistic as well as eye tracking data. Submitted to *Bilingualism: Language and Cognition* (special issue ed. by Scott Jarvis).

Klein, Wolfgang
    1994   *Time in Language.* London: Routledge.

Leclerq, Pascale
    2008   L'influence de la langue maternelle chez les apprenants adultes quasi-bilingues dans une tâche contrainte de verbalisation. *AILE*, 51–70.

Natale, Silvia
    2008   Semantische Gebrauchsdeterminanten der progressiven Verbalperiphrase stare+gerundio. Eine datenbasierte Studie. Dissertation, University of Heidelberg, Germany.

van Pottelberge, Jeroen
    2004   *Der am-progressiv.* Tübingen: Narr.

Rothstein, Susan (ed.)
    2008   *Theoretical and Crosslinguistic Approaches to the Semantics of Aspect.* Amsterdam: Benjamins.

Smith, Carlota
    1997   *The Parameter of Aspect.* Dordrecht: Kluwer. 2nd edition.

von Stutterheim, Christiane
  1997   *Einige Prinzipien des Textaufbaus*. Tübingen: Niemeyer.
von Stutterheim, Christiane and Mary Carroll
  2006   The impact of grammatical temporal categories on ultimate attainment in L2 learning. In *Educating for Advanced Foreign Language Capacities*, H. Byrnes, H. Weger-Guntharp and K. Sprang (eds.), 40–53. Georgetown: Georgetown University Press.
von Stutterheim, Christiane, Dorothea Bastin, Mary Carroll, Monique Flecken and Barbara Schmiedtová
  2008   How grammaticized concepts shape event conceptualization in the early phases of language production: Insights from linguistic analysis, eye tracking data and memory performance. Submitted to *Cognition*.
Williams, Christopher
  2002   *Non-progressive and progressive aspect in English*. Bari: Schena.
Xiao, Richard and Tony McEnery
  2004   *Aspect in Mandarin Chinese*. Amsterdam: Benjamins.

Verb aspect and the mental representation of situations

Carol J. Madden and Todd R. Ferretti

> For God, history is a landscape of events. For Him, nothing really follows sequentially since everything is co-present.
>
> Virilio (2000)

In contrast to the perspective alluded to above, human beings are unable to process situations in an atemporal manner. Indeed, every situation occurs at a certain time and endures for a certain duration. Just as it is impossible to experience situations without a temporal perspective, reference to time in language is likewise inescapable. Any situation description, no matter how simple, provides temporal information. The manner in which the temporal structure of situations is coded in language corresponds to the multitude of possible situation sequences that occur in our everyday experiences. One can make a phone call right now, tomorrow, yesterday, every Tuesday, while taking a walk, before eating breakfast, for three hours, etc. Accordingly, language offers tense markers on verbs (*called, will call*) as well as adverbial markers (*yesterday, after lunch*) that cue comprehenders to represent a situation as occurring before, during, or after some point on a timeline. Tense markers locate situations relative to the time of utterance, whereas adverbials and sequential ordering of situation descriptions provide cues for situating situations relative to each other on a narrative timeline.

Despite the fact that all situations are concisely coded in language through sequential, discrete phrases, situations as they are experienced in the real world are not so neatly packaged. Instead, they have varying duration, do not always occur in continuous sequence, and do not always wait for the previous situation to finish before beginning, which can make the ordering of situations quite complicated. For instance, reading a book takes a considerable amount of time, whereas opening a book is finished as soon as it begins. Washing a pair of jeans is a situation with multiple stages that progress toward a clear and natural endpoint, but wearing a pair of jeans is a homogeneous situation that endures indefinitely, until it is discontinued. The temporal variation found between and within situation types is of course directly related to how the participant(s) and objects of each situation contribute to the ongoing and completed development of those situations. This close relationship between the temporal and causal structure of situations is crucial for specifying how they are organized. These characteristics of the structure of situations must therefore be adequately captured in language if successful understanding of situation descriptions during language comprehension is expected to ensue. The present chapter examines how verb aspect temporally references situations as ongoing versus completed thereby influencing the mental representation of situations.

The chapter begins by introducing the characteristic properties of verb classes and aspectual categories, as well as how they interact to produce the temporal properties of sentences. Following this, problems associated with the classification schemes are discussed, and possible lines of improved analysis are mentioned. This theoretical discussion is followed by an overview of empirical research on grammatical verb aspect within the field of cognitive psychology. Recent psycholinguistic studies examining the role of aspect in constraining discourse model construction are reviewed. Finally, special issues involving verb aspect's interaction with situation types and tense are discussed, and future research directions that would contribute to the psychological relevance of aspectual distinctions are proposed.

Verb classes

A number of different philosophers and linguists have proposed verb classification schemes (e.g., Dowty 1979; Lyons 1977; Mourelatos 1981; Nordenfelt 1977; Vendler 1967; Verkuyl 1972, 1993; see also chapter 2, section 4 of this book), and it its generally agreed that verbs denote situations that fall into one of the following three core classes; states (e.g., *know*), events (*arrest*), and processes (*walk*). In addition to these classes, a fourth class, called accomplishments (e.g., *build*), is identified as having properties of both processes and events, and thus considered a complex class or compound class (Dowty 1979). These classes are found in all languages and are considered by many to be ontological categories (Kiel 1979; Lyons 1977).

Although these classes are generally agreed upon, there is far less agreement as to the composition of the classes. The disagreement among linguists arises from trying to categorize verbs into discrete classes on the basis of the presence or absence of particular temporal properties. Table 1 displays the semantic dimensions of the different verb classes. The main problem with this approach is that the meaning of individual verbs actually overlaps with the properties of multiple verb classes. The following discussion illustrates the overlapping nature of the classes.

Table 1. Semantic dimensions of the different verb classes.

Semantic Dimensions	States	Events (achievements)	Processes	Accomplishments (process+event)
Exist	+	−	−	−
Happen	−	+	+	+
Durative	+	−	+	+
Momentary	−	+	−	−
Homogenous	+	−	+	−
Heterogenous	−	+	+	+
Progressive	−	−	+	+
Cyclical	−	−	+	−
Telic	−	+	−	+
Atelic	+	−	+	−

The most basic semantic distinctions identified by linguists as distinguishing between the verb classes are that between situations that exist versus those that happen, those that are continuous or durative (i.e., extended in time) versus punctual or momentary, between those that are homogenous or unchanging versus heterogeneous and changing, and finally the distinction between telic and atelic situations (Comrie 1976; Dowty 1979; Lyons 1977). Telic situations (e.g., *build*) are those situations that are directed toward a goal, and when the goal is reached, a change of state occurs and the situation is complete. Atelic situations are those that do not have a natural endpoint (Comrie 1976).

States are the simplest of the situation types and are characterized as existing, are homogenous or unchanging throughout their duration, and are continuous because they do not have endpoints. The lack of a natural final endpoint makes states atelic. Non-stative or dynamical situations (*events* and *processes*) are characterized as happening since they involve change. If there is a single change of state, the situation is considered momentary or punctual and is called an *event* (also known as *achievements*). If an event is under control of an agent, the event is called an *act* (Lyons 1977). Because events involve a natural endpoint (i.e., the end of a change of state), but do not have a process leading up to the eventual change in state (i.e., they are punctual), they are considered to be pseudo-telic. If the situation is extended in time (i.e., durative or continuous) and includes multiple changes of state the situation is considered a *process*. However, because processes are continuous they do not have an intrinsic endpoint and are therefore considered atelic. Processes can be progressive (e.g., *lift*) or cyclical (*walk*), and if they are under control of an agent they are called *activities* (Lyons 1977). Finally, *accomplishments* such as "build", are special in the sense that they include properties of both processes and events. For example, "build" involves a process that leads up to an intrinsic endpoint (i.e., something is built). The presence of an intrinsic endpoint makes accomplishments telic in nature.

The foregoing discussion indicates that the verb classes all have features in common with one another, and therefore, the categorization of the situation types is not "clear cut." To summarize, both events and processes are differentiated from states because they happen (i.e., involve changes of state). Processes are distinguished from events and are similar to states because they are extended in time and can be homogenous. Events are similar to states because they both lack an internal phase. Finally, accomplishments share features with processes and states because they involve duration, and also with events because they involve a natural endpoint.

Aspectual categories

The grammatical category of aspect captures the different ways language refers to the temporal structure of situations through grammaticalization in the morphology (cf. chapter 2, section 3 of this book). The linguistic analysis of aspect is complicated by the fact that the languages of the world are extremely varied in how aspectual distinctions are grammaticalized and, as a result, there are many different aspectual classification schemes (Comrie 1976; Lyons 1977). From a language independent perspective, there appear to be three aspectual categories that are grammaticalized to various degrees in the world's languages, and these include the perfective (also called the *completive* by many linguists), imperfective, and perfect aspects (Johnson 1981; Lyons 1977).

Perfective aspect is differentiated from the other two categories by referencing the whole event and the different phases or parts that make up an event are ignored. Thus, the situations they denote are often considered punctual or bounded. When events and processes are used with the perfective they are viewed as a whole, or bounded, while states are viewed as unbounded, as well as bounded because their temporal properties do not include endpoints (Smith 1991).

The *imperfective aspect* makes specific reference to the internal structure of situations by focusing on their ongoing development, but does not make reference to their terminative phases. The imperfective category is frequently subdivided by linguists into *progressive* and *habitual* categories. The English progressive has received a lot of attention in the linguistic literature (e.g., Comrie 1976; Dowty 1979; Langacker 1982; Vlach 1981; Zhang 1995). This form of verb aspect is marked with the verbal form be - ing (e.g., is scoring, was scoring, will be scoring). The progressive form focuses on the internal phases of a dynamical situation (e.g., 1), and does not always occur with verbs expressing states (2).

(1) Dave was climbing a tree. (Dynamical situation)

(2) *Dave was knowing the answer. (Stative)

However, some stative sentences can be in the progressive form, and when this happens it has the effect of changing the state into an event (see *aspectual coercion* below). For instance, sentence (3) describes a stative verb in the progressive that is given an event interpretation, while in (4) it is not (Zhang 1995).

(3) They were living in Texas. (state → event)

(4) They lived in Texas. (state)

Events can also appear in the progressive form (e.g., 5), and when this happens the focus is on the preliminary phases of the event with no information as to its outcome (Smith 1991). Thus events used in a progressive form are much like accomplishments as they denote stages prior to the terminative stage of the event (i.e., phases leading up to, but not including the actual "winning" of the race). However, some verbs denoting events are odd in the progressive because it is hard to imagine them having preliminary phases, such as "finding" in (6).

(5) Dave was winning the race.

(6) ? Dave was finding his watch.

The habitual form of the imperfective aspect differs from the progressive form in that it refers to a series of a particular event, or the iteration of a process several times (Comrie 1976; Smith 1991). For example, in the following sentence the process of playing hockey is iterated several times.

(7) Dave used to be playing hockey.

The *perfect aspect* refers to times later than the situation and places emphasis on the resultant phases (or states) of situations rather than on inceptive phases (Comrie 1976). In English this aspectual category is marked with <u>to have + past participle</u> (e.g., has scored, had scored, will have scored). In general, the perfect functions to indicate the continuing relevance of a past situation into the present or to other reference times (Comrie 1976). The perfect aspect appears with all situation types.

The above discussion on the verb classes and the aspectual categories was purposefully laid out in a manner that would illustrate some of the problems that linguistic theories of aspectuality have to overcome. Two problems that are immediately apparent for linguistic theories is how to account for lack of clear boundaries between the verb classes, and how to explain why members from the same verb class interact variably with aspectual categories (e.g., "winning" combines more readily than "finding" in the progressive form of the imperfective). These problems are difficult to handle by linguistic theories that try to partition classes into discrete categories on the basis of the presence or absence of particular properties (e.g., Mourelatos 1981; Vendler 1967), or try to account for the classes on the bases of rules that govern aspectual composition, such as model-theoretic semantics (e.g., Dowty 1979).

A more appropriate line of analyses that can handle these difficulties is to employ a prototype theory of aspectuality (e.g., Zhang 1995). From this perspective, verb classes are not defined by strict criteria that are necessary and sufficient for defining membership, but rather their membership can be a matter of degree (Rosh 1978). This view of using a prototype is consistent with the notions of other cognitive linguists (e.g., Langacker 1982, 1987; Talmy 1988) who suggest that aspectual classes emerge from a child's interaction with the entities and things in their environment, and it is this interaction that allows the child to develop the conceptual basis for classifying the situations that occur in the world. In this view, our knowledge of the world is embodied, that is, rather than existing independently in the world, conceptual knowledge is grounded in bodily movement, perception, and all of our social and physical experiences (Lakoff 1987). When viewed from a cognitive view point, the "fuzzy" boundaries between aspectual notions is not unexpected.

Another line of cognitive linguistic analysis that seems to be promising with respect to solving these problems has been proposed by Narayanan and colleagues (Chang, Gildea & Narayanan 1998; Narayanan 1997). In this view, the compositionality of verb classes are grounded in sensory-motor control primitives such as goal, periodicity, iteration, final state, duration, and other parameters such as force and effort. Aspectual expressions are considered to be "linguistic devices that refer to schematized processes that recur in sensory-motor control". The schematized properties have been implemented in a computational model that provides a semantic representation of aspect that is grounded in the execution of the model. The semantic representation of aspect naturally falls out of the executing model without the need for special interpretation rules (such as those proposed by model theoretic models). The viability of this model is demonstrated by its ability to simulate a number of important characteristics of aspect that have traditionally been problematic, such as accounting for cross-linguistic variation in aspectual expressions, verbs that are difficult to classify, and the interaction between verb class and aspectual categories.

Empirical investigations of verb aspect

The use of verb aspect in discourse provides comprehenders with a good deal of temporal information about the described situations. Exactly how verb aspect constrains representations of described situations has traditionally been the focus of linguistic and philosophical discussions, such as many of those cited above, but recently this topic has also become the focus of a growing number of empirical investigations in the area of cognitive psychology. This line of research has shown that verb aspect provides two types of information about the situations it is used to describe. Specifically, verb aspect has been shown to provide information about a situation's completion status, as well as information about the availability of people, entities, and features of the described situation. These two sets of findings will be summarized and discussed below.

The mental representation of ongoing versus completed situations

There have been several investigations of how verb aspect serves to constrain comprehenders' representations of situations with respect to completion status. In general, these studies have considered the distinction between the perfect or perfective (completive) aspects and the imperfective aspect. As described above, linguists have argued that the perfect aspect emphasizes the resultant rather than inceptive phases of situations, and the perfective aspect describes situations as complete wholes, without focus on internal phases or parts that make up the situations. As the perfect aspect focuses on resultant phases, and the perfective aspect collapses across the phases of a situation, treating it as an indivisible unit, both aspectual categories should yield representations of completed situations. Therefore, these aspectual categories are often compared to the imperfective aspect, which makes specific reference to the internal phases of situations while ignoring their terminative phases. Indeed, several studies in the arena of cognitive psychology have shown that the imperfective aspect focuses representations on the ongoing development of situations, whereas the perfective and perfect aspects focus representations on the endpoint or resulting phases of situations.

For instance, Magliano and Schleich (2000) have demonstrated that situations described in the imperfective aspect are comprehended as still ongoing, whereas situations described in the perfective aspect are more often understood as completed. In a set of 4 experiments, they presented comprehenders with stories in which a critical situation was described either in the imperfective or perfective aspect (i.e., was changing a tire vs. changed a tire). This critical sentence was followed by three sentences describing situations that could be understood to occur either during or after the critical situation. In Experiment 1, participants were asked a question regarding whether or not the critical situation was ongoing or completed at four positions in the story; after the critical sentence and after each of the three subsequent sentences. Across the four question positions, participants were more likely to respond that the critical situation was still ongoing when the imperfective aspect had been used rather than the perfective aspect.

Experiment 2 employed the same design as Experiment 1 with the addition of a manipulation of the typical duration of the critical situation (i.e., was writing/wrote a check vs. was writing/wrote a novel). Again, participants were more likely to respond that the critical situation was still ongoing when the imperfective aspect had been used rather than the perfective aspect. Only in the imperfective condition did typical duration have an effect, in that short duration situations were more likely to be judged as completed than long duration situations, although this effect was only observed at the late question position (three sentences after the critical sentence). These two experiments demonstrate that the use of verb aspect indeed constrains situation representations with respect to completion status. Moreover, Experiment 2 also demonstrated how comprehenders more often used the linguistic construction of verb aspect over world knowledge about the typical duration of situations to determine whether a situation was ongoing or completed.

Morrow (1985) also provides elegant demonstrations of how verb aspect constrains the temporal focus of situations. In these experiments, participants first memorized the layout of a model house and then read short narratives describing a sequence of situations that took place in that house. Across experiments, Morrow found that participants were more likely to locate target references in the path room rather than the goal room after reading sentences in the progressive past such as "She was walking past the study to the bedroom", whereas the opposite was true after reading sentences in the simple past such as "She walked past the study to the bedroom." This effect was clearly measurable over and above effects of prepositions (e.g., from/through/past, to/into) as well as order of mention of source and goal rooms (e.g., from the study to the bedroom, vs. to the bedroom from the study). Thus, this provides another demonstration that the imperfective aspect (progressive past) focuses representations on the ongoing development of a situation, whereas the perfective aspect (simple past) focuses representations on the endpoint or resulting phases of a situation.

Madden and Zwaan (2003) used a picture verification task to demonstrate how the perfective aspect constrains representations to the endpoint of a situation. In Experiment 1, participants read sentences either in the perfective ("The boy built a doghouse.") or imperfective aspect ("The boy was building a doghouse."), and then chose which of two pictures best matched the sentence. The two pictures illustrated the described situation either in progress (half-built doghouse) or completed (finished doghouse). Participants were more likely to choose the completed picture after having read a perfective sentence, but chose the in-progress and completed picture equally often after having read the imperfective sentences. In Experiment 2, only a single picture was presented (either in progress or completed) and participants were to judge as quickly as possible whether the picture matched the sentence. Although accuracy did not differ for the completed and in progress pictures, responses to verify that the picture matched the sentence were faster for the completed picture rather than the in progress picture after having read a perfective sentence. There was no difference in response speed for the two picture versions after having read an imperfective sentence. Experiment 3 reversed the order of presentation, so that one of the picture versions was presented first, and then participants were asked to judge whether a subsequent sentence matched the depicted situation. Once again, only on the perfective sentences were participants faster to respond if they had previously viewed the completed pictures than if they had viewed the in-progress pictures. Participants responded to imperfective sentences equally fast whether they had viewed the in-progress or completed pictures.

In this study, all three experiments showed that participants preferred pictures showing completed situations to pictures of ongoing situations upon reading perfective sentences but showed no picture preference upon reading imperfective sentences. This apparent lack of constraint in the imperfective construction does not necessarily mean that participants were not activating ongoing representations of the imperfective sentences. Rather, this lack of effect was hypothesized to arise from the fact that ongoing situations may be represented in various stages of completion, even approximating the endpoint of the situation. Therefore, it is more difficult to capture the appropriate stage of completion that would effectively match the participants' representations of the imperfective sentences in the picture stimuli. Conversely, the endpoint of a situation is temporally well-specified, and thus it is easier to pictorially capture the appropriate stage of completion that would effectively match the participants' representations of the perfective sentences.

Finally, Rohde, Kellar, and Elman (2006) have also demonstrated how verb aspect constrains the temporal focus of a situation. Participants were presented with transfer sentences followed by an ambiguous pronoun (e.g., "John handed a book to Bob. He _____") and were required to continue the second sentence. Given the ambiguity of the pronoun at the beginning of the second sentence, participant's responses could either continue that sentence with reference to John, the source of the transfer situation, or with reference to Bob, the goal of the transfer situation. When the initial sentence was presented in the imperfective aspect (was handing), participants were more likely to continue the second sentence with reference to John, but when the initial sentence was presented in the perfective aspect (handed), participants continued the second sentence with respect to John and Bob equally often. This demonstrates that verb aspect can impinge on the influence of typical subject preference (Crawley, Stevenson & Kleinman 1990) and first-mention privilege (Gernsbacher & Hargreaves 1988) in resolution of pronoun ambiguity. When the imperfective aspect is used there is a clear focus on the source of the transfer situation, but this focus shifts in favor of the goal or end state of the situation when the perfective aspect is used.

Verb aspect and the activation of participants, objects, and locations of situations

In addition to providing information about situation completion status, verb aspect has also been shown to influence the availability of people, entities, and features of described situations. For instance, Carreiras et al. (1997) tested the influence of verb aspect on the availability of characters in short narratives. In this study, participants read short passages in which the actions of two characters were described. The target character was described as doing something either in the imperfective or the perfect aspect, and subsequently, that character's name was presented as a probe word. Participants were faster to respond to the probe when the actions of that character were described in the imperfective rather than the perfect aspect. Thus the imperfective aspect acts as a cue to keep a character in the focus of the mental model for the unfolding narrative.

Ferretti and colleagues (2007) used both behavioral and neurocognitive methodology to test how verb aspect influences the availability of typical locations of situations. In their first experiment, they used a semantic priming paradigm in which participants had to read a verb phrase silently and then name a noun (location) aloud. The verb-noun pairs could be related (was skating – arena) or unrelated (was praying – arena). Furthermore, the verb could either be presented in the past imperfective (was skating, was praying) or in the past perfect (had skated, had prayed). Participants' times to name the location word showed a semantic priming advantage for related over unrelated pairs in the past imperfective, but no relatedness difference occurred for naming times of locations that followed the past perfect verbs. Thus, these results show that location information is more available when the imperfective aspect is used rather than the perfective, and that this difference in availability is evident very quickly when verbs are read.

In a second experiment involving a sentence completion task, participants generated a larger proportion of locative prepositional phrases as continuations to sentence fragments with past imperfective verbs ("The cow was grazing ____") compared to sentence fragments with past perfect verbs ("The cow had grazed ____"). Furthermore, a third experiment utilizing event-related brain potential (ERP) methodology examined online reading of sentences with high-expectancy locations ("The diver was snorkeling/had snorkeled in the ocean.") and low-expectancy locations ("The diver was snorkeling/had snorkeled in the pond."). Only the sentences in the past imperfective yielded a larger N400 (an ERP component known to index the ease of semantic integration of words in text) for low-expectancy locations than for high-expectancy locations. The past perfect sentences showed no expectancy difference in the N400. Also, slow cortical potentials that developed over the prepositional phrases were more negative following perfect than imperfective verbs, indicating that locative prepositional phrases following perfect verbs were more difficult to integrate. Finally, early sensory ERP components known to be sensitive to visual processing and visual selective attention (e.g., P1, P2) also varied systematically as a function of aspect and expectancy. Specifically, within 200 ms of the onset of location nouns, the amplitudes of these components were most reduced for typical locations following perfect verbs relative to all other conditions. Although these early ERP component results need further investigation, they suggest that verb aspect acts as a cue to actively suppress location information following perfect verbs.

Not only do characters and locations become more accessible when the imperfective aspect is used rather than the perfect or perfective aspects, but also typical instruments such as hammers and knives are more active when their corresponding situations are described in the imperfective rather than the perfect aspect. Truitt and Zwaan (1997) presented short texts in which one of the described actions implied the use of a specific instrument, such as a hammer. Participants were faster to verify that the instrument had been mentioned in the preceding text after reading the target action sentence in the imperfective aspect ("Jason began pounding the nails into the board.") rather than in the perfective aspect ("Jason pounded the nails into the board."). Likewise, when the instrument had not been mentioned in the previous text, participants were slower to reject the probe word "hammer" after reading the imperfective than the perfective version of the target action sentence. These findings demonstrate that instruments are more available in a reader's mental model when situations are described as ongoing (imperfective) rather than completed (perfective).

Finally, we revisit Magliano and Schleich's (2000) study as it also provides evidence that verb aspect can influence the time course of overall activation of a situation. Specifically, this study demonstrates that the imperfective aspect increases activation for an entire situation more so than the perfective aspect. In their third experiment, participants read texts that introduced a target activity either in the imperfective or the perfective aspect. At two later points in the story a probe verb phrase that was related to the target activity was presented (change tire), and participants were required to judge whether the verb phrase corresponded to one of the activities mentioned in the story. Across time points, the probe verb phrases were responded to more quickly following imperfective activity descriptions than perfective activity descriptions. Furthermore, their fourth experiment reveals that high span comprehenders are better able than low span comprehenders to maintain activation of activities described in the imperfective aspect. This finding provides evidence that the imperfective aspect acts as a cue to maintain activation of a situation, whereas situations described in the perfective aspect are allowed to become less active.

Taken together, empirical investigations on this topic have shown that verb aspect has an important role in language comprehension. Verb aspect provides cues in composing the temporal structure as well as the focus of situation representations. The preceding section has discussed several demonstrations that situations described in the imperfective aspect are represented as ongoing while situations described in the perfect or perfective are represented as completed (Madden & Zwaan 2003; Magliano & Schleich 2000; Morrow 1985; Rohde et al. 2006). Subsequently, we discussed studies showing that situations described in the imperfective rather than the perfect and perfective aspects lead to greater and longer activation of the situations as well as their features (Carreiras et al. 1997; Ferretti et al. 2007; Magliano & Schleich 2000; Truitt & Zwaan 1997). While this discussion has provided insight on how aspect constrains the mental representation of situations in general, the following section will take a closer look at how aspect differentially affects the representation of various classes of verbs.

The interaction of aspect and verb classes to constrain mental representation

As previously discussed in the beginning of this chapter, the semantic properties of the different verb classes (Events, Processes, Accomplishments, States) lead to differences in how the classes interact with aspectual morphemes. In the following section we explore the notion that interactions between the verb classes and aspect have consequences on how mental representations of situations are constructed during language comprehension. There has been relatively little direct empirical investigation of how aspect and verb classes combine to constrain the construction of mental representations of situations in the cognitive and psycholinguistic literature. In most cases, research in these fields involved sets of verbs that came from a single verb class (e.g., Madden & Zwaan 2003; Morrow 1985), or involved multiple verb classes without the purpose of directly contrasting the classes (e.g., Ferretti et al. 2007).

Recall that one of the main differences between accomplishments and processes is that the later do not have a natural endpoint and are thus considered atelic. As a result, using aspect to denote processes as completed, and thus with an endpoint, may lead to greater difficulty constructing mental representation of those situations relative to completed accomplishments which have lexical semantics that include natural endpoints. Thus, there may be a perfective advantage in terms of processing costs for accomplishments and, alternatively, an imperfective advantage for processes. A recent study by Yap, Kwan, Yiu, Chu, Wong, Matthews, & Shirai (2006) investigated this possibility using a forced-choice picture matching task similar to Madden and Zwaan (2003). Across two experiments, native Cantonese speaking participants were presented with auditory sentences that either contained activity or accomplishment verbs marked with perfective and imperfective aspect. Immediately following these sentences, two pictures were displayed on a computer screen. One of these pictures depicted the situation mentioned in the auditory sentences as complete, whereas the other picture depicted the same situation in progress. Participants indicated as quickly as possible which picture was the best match for the situations described in the auditory sentences. The results demonstrated that verb aspect had a different influence on responses to accomplishment and activity verbs. For accomplishment verbs, people took significantly longer to select the matching pictures when they depicted ongoing versus completed situations. On the other hand, matching pictures for activity verbs were verified more slowly when they depicted completed situations. These results show that forcing completion status on situations that do not have natural endpoints, or forcing ongoing status to situations that have well-defined endpoints, comes with a processing cost.

Often in cases like these, when the grammatical aspect of a verb is at odds with that verb's inherent temporal semantics, the semantic system employs a function called aspectual coercion. Aspectual coercion refers to the forcing of an alternative interpretation of a particular situation when the usual interpretation does not fit with the grammatical constraints of the sentence context (Moens & Steedman 1988). For instance, the inherent temporal semantics of the verb *sleep* are intact when we say that a child slept for three hours, because the process of sleeping can endure for three hours. However, the inherent temporal semantics of the achievement verb *jump* are incompatible with a duration of three hours, so when we say that a child jumped for three hours, the alternative interpretation of iterative jumping is forced to supplant the usual single jump.

While little empirical evidence has addressed the topic of aspectual coercion, a couple of studies have shed some light on the representational underpinnings of this process (Piñango, Winnick, Ullah, & Zurif 2006; Piñango, Zurif, & Jackendoff 1999). Piñango and colleagues showed that the function of aspectual coercion was associated with a processing cost during sentence comprehension. Participants heard sentences that did or did not require aspectual coercion (e.g., *The insect hopped/glided effortlessly until it reached the far end of the garden that was hidden in the shade.*). In this example, hopping is an achievement that is over almost as soon as it begins, whereas gliding is a process that can endure for a specified period of time, such as indicated here with the temporal modifier, "until". Thus, hopping requires an alternative interpretation, whereas gliding can be interpreted in a straightforward way. While participants heard these sentences, they were also required to make a lexical decision to a word presented on the screen, a secondary task that is assumed to compete for resources with the primary task of sentence comprehension. The lexical decision word was presented 250 ms after hearing the temporal modifier "until," which was presumed to be when comprehenders would be engaged in the semantic selection process for the temporal interpretations of the gliding/hopping situations.

Indeed, Piñango and colleagues observed significantly longer lexical decision times when the concurrent sentence required aspectual coercion than when the concurrent sentence afforded the straightforward interpretation of the verb (Piñango et al. 2006; Piñango et al. 1999). Longer lexical decision times were not observed when the lexical decision had to be made 250 ms earlier, at the point when the constraining syntactic information is presented, suggesting that this semantic selection process of aspectual coercion operates at a delay (Piñango et al. 2006). Todorova, Straub, Badecker, & Frank (2000) further demonstrated that this processing cost is associated with the operation of aspectual coercion rather than merely the representation of iterativity in general.

The verb classes discussed in this chapter capture the main differences between all of the situation types. However, within these types of situations there are of course groups of verbs that overlap semantically in more specific ways than the properties outlined in Table 1. Due to the large variability in the temporal and causal structure of the situation types, it should not be surprising that several different classes have been identified and shown by linguists to have relatively specific syntactic behavior as a result of their lexical semantic properties (see Levin 1993 for a discussion of many of these more specific verb classes and their syntactic alternations). To date however, there has been little empirical research on how these more specific verb classes interact with the morphosyntactic properties of verb aspect to constrain mental representation of situations. To our knowledge, only the recent research by Rohde et al. (2006) has directly examined how verb aspect interacts with the lexical semantic properties of more specific verb classes. Recall that participants in this study read sentences that included verbs of transfer (*give, send*) that were in either imperfective or perfect form. They were then provided with a pronoun that was ambiguous with respect to whether it referred to the source participant or goal participant in the preceding transfer situation. Participants were required to generate sentence continuations that seemed natural to them (*Jim gave the book to Tom.*

He_____). Rohde et al. also examined how specific semantic differences within verbs of transfer, such as whether the participants are co-located (e.g., co-located – *give*; not co-located – *mail*) and whether the verbs guarantee transfer (*give*) or not (*mail*), interact with verb aspect to constrain sentence completions. Their results demonstrated similarities and differences in how these specific lexical properties influence completions. For example, across all three of their conditions (co-located, guaranteed transfer/co-located, no guaranteed transfer/not co-located, no guaranteed transfer), they found that people were more likely to generate completions in which the ambiguous pronoun referred to the source rather than the goal participants in the preceding sentences, but only when imperfective aspect was used. When the sentences were in perfective form they also found a greater number of source completions, but this was only true for verbs of transfer that did not involve co-location of participants and did not guarantee transfer. The other two verb classes had approximately equal source and goal continuations. Thus, differences in co-location of participants between verbs of the same broad verb class (i.e., verbs of transfer) can lead to differences in how people focus on participants mentioned in situations and thus how people resolve the resolution of ambiguous pronouns in discourse.

The foregoing discussion indicates that more general lexical semantic features that distinguish the main verb classes (e.g., telic vs atelic), as well as more specific semantic features that comprises sets of verbs within these broad classes (e.g., co-location of participants), have important consequences for how verbs combine with verb aspect morphemes. Understanding how these general and specific semantic features associated with the situation types interact with and are coerced by verb aspect is crucial to understanding how people construct the mental representation of situations during language comprehension. Thus, it is important that future research on verb aspect is conducted to identify which verb-specific semantic features are critical for constraining the way aspect interacts with the various situation types. Finally, it should also be noted that it is critical that researchers are in general aware that such differences in the lexical semantic properties between and within verb classes will influence the pattern of results obtained in their experiments.

The interaction of aspect and tense to constrain mental representation

At the outset of the chapter, we briefly discussed how situations are located with respect to the time of utterance (most often now). This temporal information is coded in language through tense operators. The past tense locates situations earlier than the time of utterance, the present tense locates situations during the time of utterance, and the future tense locates situations later than the time of utterance. Most research on verb aspect and situation representations has been conducted within the past tense, and this is for several reasons. First, much of our discourse about situations concerns accounts of what has occurred prior to the time of utterance, and the past tense is the natural manner of discussing these situations. Therefore, the past tense is an ecologically valid setting for research on situation representations. Second, and more important, is the fact that verb aspect behaves differently within the different tenses. Because of the diversity of possible internal situation structures, some situations are accommodated quite easily in the past, present, and future tenses whereas other situation structures are not. Likewise, some grammatical aspects occur within certain tenses with straightforward interpretations, while other grammatical aspects force unusual interpretations in some tenses. These tense-aspect interactions will be discussed below.

In the present tense, situations are anchored to the time of utterance, which normally corresponds to the current now. However, whereas the past and future are durative regions of time, the current now is in fact a point in time without duration. This creates problems for any situation that has duration, as a durative situation cannot be accommodated within a non-durative point in time. When an infinitely small point in time like the present attempts to encapsulate a situation that takes time, such as drinking coffee, representational problems will naturally occur. In this case, alternative interpretations are forced. As discussed in the previous section, an alternative interpretation of a particular situation can be coerced when the situational aspect and the grammatical aspect of that situation are at odds (Moens & Steedman 1988; Piñango et al. 2006; Piñango et al. 1999). Aspectual coercion also occurs when the situational aspect of a verb and the tense of the sentence are at odds, as is often the case for the present tense.

Because the progressive present tense (i.e., *am drinking*) describes a situation from an imperfective perspective in which the onset of the situation has already occurred and the endpoint has not been reached, situations with duration can be accommodated in the present tense when this construction is used. The situation is represented as ongoing at the current point in time. However, the simple present tense is especially sensitive to aspect, as it describes a situation from a perfective perspective in which the entire situation, including onset and endpoint are collapsed into a complete whole. Thus, when processes or events are presented in the simple present tense, they are not interpreted in a straightforward manner, but rather coerced to yield alternative interpretations. These include iterative or habitual interpretations, or idiosyncratic interpretations such as the historical present used in news reports and joke telling. For instance, when we use the simple past tense to say, *Peter parked Susan's car*, we assume that this happened once. But when we use the simple present, as in *Peter parks Susan's car*, then a habitual interpretation is assumed in which Peter often parks Susan's car for her. Likewise, the simple present is often used to recount situations that occurred in the past as if they are being experienced now, as in newspaper headlines (*Mayor funds new recreation center*") and joke telling (*A man walks into a bar*.). These alternative interpretations of situations are forced because of the incompatibility of the point in time that is the present tense, and the normal duration of these situations. Although language users have become quite adept at automatically re-interpreting situations when aspectual coercion is required, this process nonetheless incurs a processing cost (Piñango et al. 2006; Piñango et al. 1999).

Not all durative situations are problematic for the simple present tense. A state (*being sick, having a car*) is temporally unbounded, yet is easily accommodated in the simple present tense (*John is sick, I have a car*). This is because states are homogeneous and have no dynamic properties. They can be sampled at any point in time, always yielding the same evaluation. Thus, the state of being sick or having a car can be true at the current point in time and likewise produce straightforward interpretations when described in the simple present tense (perfective). However, as mentioned in the previous section, states do not lend themselves to the imperfective aspect in any tense in English (*John was/is/will be having a car*). Even though the idea of an unbounded state seems to fit nicely with the idea of an ongoing situation, and indeed states are described in the imperfective in many other languages, states are not described using the imperfective aspect in the English tenses. Because the English imperfective is employed to make an event more like a state (the state of an event as ongoing and continuous), this construction is not compatible with situations that are already inherently stative.

A further complication with aspectual distinctions in the present tense is that the present perfect (*John has eaten*) is often wrongly categorized as a past tense. This is a result of the fact that its grammatical aspect refers to a completed situation. However, this construction is technically categorized as a present tense, because it refers to the resultant state of that completed situation and its relevance for the present situation. This becomes clear when speaking of a deceased person, as we cannot say that he has eaten.

The present auxiliary (have) implies relevance of having eaten for the present situation, and thus demands that the agent be alive and present.

This dynamic interaction between tense and aspect is especially difficult for young children to grasp, and not only in the tricky case of the present perfect. In fact, some theorists argue that very young children are generally unable to correctly map both forms of grammatical information onto the situations they describe. The *Aspect First Hypothesis* claims that during the early stages of language learning, children first understand whether a situation is completed or not, rather than whether it occurred in the past, present, or future (Wagner 2001). According to this hypothesis, children code for aspect rather than tense in their verbal morphology during this stage. Wagner (2001) provides evidence for this claim in a study in which two and three year old children watched a toy cat perform various actions, such as drawing, jumping, or emptying a container, multiple times along a path. The children were then asked to point where the cat is/was jumping. In the case that actions were completed in the past (a complete face was drawn and a container was emptied entirely) but incomplete in the present (the question was asked mid-action), the aspect and the tense information were complimentary, and children responded correctly to the questions with regard to tense. However, when the children watched the cat perform a sequence of incomplete actions (only a half face was drawn, or some contents of the container remained) then the youngest children had difficulty making decisions on the tense questions. At the very least, this demonstrates that children sometimes use aspectual information about a situation's completion status at the expense of tense information about the temporal ordering of situations. It is argued that this bias for aspectual information arises because the completion status of a given situation is a more basic construct for children to understand than the temporal ordering of several situations.

The current discussion demonstrates how tense and aspect interact to produce dynamic perspectives on the semantic representations of situations. Indeed there are complications that arise during this process, as the two types of grammatical information do not always accommodate each other very well. In these cases, alternative interpretations of situations are coerced so that the temporal structure of the situation can be accommodated within a given tense. Furthermore, we have also seen that in the case of very young children, tense information can even be ignored. Understanding how the temporal structure of situations (verb aspect) interacts with and is coerced by the temporal ordering of situations (tense) provides important insights into mental representations of situations. While most research on verb aspect has been conducted in the past tense, it is important that more research is conducted in the present and future tenses to better understand the nature of the tense-aspect interaction during language comprehension.

Conclusions

The present chapter presents an overview of how the topic of verb aspect has been addressed in the domain of cognitive psychology. We began by detailing the traditional classification schemes of temporal situation structure, as well as describing how the verbs that describe these situations fall into the separate aspectual classes. In doing so, several shortcomings of the current classification schemes were revealed, and innovations in addressing these shortcomings were cited. A potential solution resides in prototype theories of aspect that emerge from actual experience in the environment and allow degrees of membership within aspectual categories. In addition, recent theories that ground the verb classes in sensory-motor control primitives such as goal, periodicity, iteration, final state, duration, and force offer promising innovations for the system of aspectual classification. These novel classification schemes provide a fruitful area for future research, as it is important for researchers to empirically validate the representational underpinnings of the aspectual classifications.

In subsequent sections, we reviewed empirical research on grammatical and situational verb aspect within the field of cognitive psychology. A discussion of recent studies revealed that verb aspect yields two important sources of information during language comprehension. First, several studies have demonstrated how verb aspect provides information about a situation's completion status, whereby the perfect and perfective aspect yield a representation of a completed situation, and the imperfective aspect yields a representation of the situation as ongoing. Second, verb aspect provides information about the availability of people, entities, and features of the described situation. Several studies show that properties of situations such as agents, instruments, and locations are more available when the situations are described in the imperfective aspect rather than the perfect or perfective aspects. While these studies offer much information about the representation of a situation described in the imperfective, it remains unclear exactly how perfect and perfective situations are represented. It has been shown that the perfect and perfective aspects are represented as completed or complete situations, but this conception requires further clarification in order to provide a better understanding of the construction of mental models during language comprehension.

Finally, we discussed special issues involving the interaction between verb aspect and various situation types, as well as the interaction between verb aspect and tense. Of special interest in these sections is the process of aspectual coercion, in which an alternative interpretation of a situation is forced under conditions wherein the inherent aspectual characteristics of a verb do not mesh with its tense or grammatical aspect. Although there have been some preliminary investigations of this issue (Piñango et al. 2006; Piñango et al. 1999; Todorova et al. 2000; Wagner 2001), more research is required to understand how inconsistencies in the various sources of temporal information are identified and resolved during language comprehension.

In closing, it should be restated that the aspectual distinctions and characteristics discussed in this chapter are not only relevant for theories of verb classification and the temporal representation of situations. Surely, these grammatical and semantic distinctions also stand to influence the pattern of results obtained in experiments in all areas of language and situation comprehension. Thus, researchers in any domain that employs the presentation of verbs should be aware of the aspectual issues raised in the present chapter.

References

Carreiras, Manuel, Nuria Carriedo, Maria Ángeles Alonso and Ángel Fernández
    1997   The role of verb tense and verb aspect in the foregrounding of information during reading. *Memory & Cognition* 25: 438–446.

Chang, Nancy, Daniel Gildea and Srini Narayanan
    1998   A Dynamic Model of Aspectual Composition. In *Proceedings of the 20th Cognitive Science Society Conference*. 226–231. Madison: Lawrence Erlbaum

Comrie, Bernard
    1976   *Aspect: An introduction to the study of verbal aspect and related problems*. New York: Cambridge University Press.

Crawley, Rosalind A., Rosemary J. Stevenson and David Kleinman
    1990   The use of heuristic strategies in the interpretation of pronouns. *Journal of Psycholinguistic Research* 4: 245–264.

Dowty, David R.
    1979   *Word meaning and montague grammar*. Reidel: Boston.

Ferretti, Todd R., Marta Kutas and Ken McRae
    2007   Verb aspect and the activation of event knowledge. *Journal of Experimental Psychology: Learning, Memory, and Cognition* 33: 182–196.

Gernsbacher, Morton A. and David J. Hargreaves
 1988 Accessing sentence participants: The advantage of first mention. *Journal of Memory and Language* 27: 699–717.

Johnson, Marion R.
 1981 A Unified Temporal Theory of Tense and Aspect. In *Syntax and Semantics: Tense and Aspect* 14: 145–175. Toronto: Academic Press.

Keil, Frank C.
 1979 *Semantic and Conceptual Development: an ontological perspective.* Cambridge, MA: Harvard University Press.

Lakoff, George
 1987 *Women, fire, and dangerous things: What categories reveal about the mind.* Chicago, IL: University of Chicago Press.

Langacker, Ronald W.
 1982 Remarks on English aspect. In *Tense/aspect: Between semantics and pragmatics*, P. Hopper (ed.). Amsterdam: John Benjamins.
 1987 *Foundations of cognitive grammar, Vol. 1: Theoretical Prerequisites.* Stanford, CA: Stanford University Press.

Levin, Beth
 1993 *English Verb Classes and Alternations: A Preliminary Investigation.* Chicago, IL: University of Chicago Press

Lyons, John
 1977 *Semantics. Vol. 2.* Cambridge: Cambridge University Press.

Madden, Carol J. and Rolf A. Zwaan
 2003 How Does Verb Aspect Constrain Event Representations? *Memory & Cognition* 31: 663–672.

Magliano, Joseph P. and Michelle C. Schleich
 2000 Verb aspect and situation models. *Discourse Processes* 29: 83–112.

Moens, Marc and Mark Steedman
 1988 Temporal ontology and temporal reference. *Computational Linguistics* 14: 15–28.

Morrow, Daniel G.
 1985 Prepositions and verb aspect in narrative understanding. *Journal of Memory & Language* 24: 390–404.

Mourelatos, Alexander P. D.
 1981 Events, Processes, and States. *Syntax and Semantics: Tense and Aspect* 14: 191–211. Toronto: Academic Press.

Narayanan, Srini
 1997 Talking the talk is like walking the walk: A computational model of verbal aspect. In the *Proceedings of the 19th Cognitive Science Society Conference.*

Nordenfelt, Lennart
 1977 *Events, Actions, and Ordinary Language.* Doxa Studies in the Philosophy of Language. Lund, Sweden: Bokforlaget Doxa.

Piñango, Maria M., Aaron Winnick, Rashad Ullah and Edgar Zurif
   2006   Time-course of semantic composition: The case of aspectual coercion. *Journal of Psycholinguistic Research* 35: 233–244.
Piñango, Maria M., Edgar Zurif and Ray Jackendoff
   1999   Real-time processing implications of enriched composition at the syntax-semantics interface. *Journal of Psycholinguistic Research* 28: 395–414.
Rohde, Hannah, Andrew Kehler and Jeffrey L. Elman
   2006   Event Structure and Discourse Coherence Biases in Pronoun Interpretation. In *Proceedings of the 27th Annual Meeting of the Cognitive Science Society*.
Rosh, Eleanor
   1978   Principles of categorization. In *Cognition and categorization*, E. Rosch & B. B. Loyd (eds.), 27–48. Hillsdale, NJ: Erlbaum.
Smith, Carlota S.
   1991   *The Parameter of Aspect*. Boston: Kluwer.
Talmy, Leonard
   1988   Force Dynamics in Language and Cognition. *Cognitive Science* 12: 49–100.
Todorova, Marina, Kathy Straub, William Badecker and Robert Frank
   2000   Aspectual coercion and online computation of sentential aspect. In *Proceedings of the 22nd Annual Conference of the Cognitive Science Society*, Philadelphia, PA.
Truitt, Timothy P. and Rolf A. Zwaan
   1997   Verb aspect affects the generation of instrument inferences. Paper presented at the *38th Annual Meeting of the Psychonomic Society*, Philadelphia, November 1997.
Vendler, Zeno
   1967   *Linguistics in Philosophy*. Ithaca, NY: Cornell University Press.
Verkuyl, Henk J.

   1972   *On the compositional nature of the aspects*. Dordrecht: Reidel.
   1993   *A theory of Aspectuality*. Cambridge: Cambridge University Press.
Virilio, Paul
   2000   *A Landscape of Events*. Cambridge: MIT Press.
Vlach, Frank
   1981   The semantics of the progressive. In *Syntax and Semantics: Tense and Aspect* 14: 271–292. Toronto: Academic Press.
Wagner, Laura
   2001   Aspectual influences on early tense comprehension. *Journal of Child Language* 28: 661–681.

Yap, Foong H., Stella W. M. Kwan, Emily S. M. Yiu, Patrick C. K. Chu, Stella F. Wong, Stephen Matthews, and Yasuhiro Shirai
    2006   Aspectual Asymmetries in the Mental Representation of Events: Significance of Lexical Aspect. In *Proceedings of the 28th Annual Meeting of the Cognitive Science Society*.

Zhang, Lihua
    1995   *A contrastive study of aspectuality in German, English & Chinese.* Lang: New York.

Computational modeling of the expression of time

Ping Li and Xiaowei Zhao

1. Introduction

To describe and exchange information about time is an important part of human activities. The expression of time is therefore one of the central conceptual domains of our language use (see chapter 1 of this book and Bates, Elman & Li 1994). When we are talking, we describe situations as being in the past, present or in the future, and we talk about events as ongoing or completed. There are two key linguistic categories for expressing temporal concepts in the world's languages: tense and aspect. Tense is often concerned with the chronological ordering of situations that happen at different time points, and is often used to locate the relationship between time of event and time of speech. In contrast, aspect typically characterizes how a speaker views the temporal contour of a situation described, for example, the beginning, continuation, or completion of a situation.

As temporal contours and relationships of events figure prominently in people's speech activities, it is important for any given human language to have a capable system for expressing these events and relationships, and for speakers/listeners to learn and process this system. Empirical evidence appears to support the idea that tense and aspect are among the earliest linguistic devices acquired by children, and as such the scientific study of the expression of time provides significant insights into not only how young children acquire temporal notions, but also what psycholinguistic mechanisms underlie the general acquisition processes.

In this chapter, we review computational models of the expression and acquisition of temporal concepts in language. With the advancement of modern computers in the last decades, computational modeling has become a very powerful methodology in many disciplines, including cognitive science and psycholinguistics. With respect to our focus here, computational models can help us introduce explicit, controllable and testable mechanisms to understand the linguistic phenomena related to temporal expressions. In addition, computational models often include certain levels of simplification in terms of language details, which makes them easier to study than traditional empirical studies that are often costly. The models and simplified datasets both make it possible for us to examine the underlying mechanisms of temporal concepts without getting entangled by specific details or certain noise in the linguistic input.

The computational modeling of tense has attracted much attention in psycholinguistics and cognitive science since Rumelhart and McClelland (1986) introduced a simple feed-forward neural network model[1] (the R&M model) to account for children's acquisition of English past tense. Given the prominence of their model and the subsequent debates in the field, we will only provide a very brief review on the acquisition of the English past tense. The primary focus in our discussion will be on various computational models that account for the expression and acquisition of aspect. Of course one needs to realize that the expressions of tense and aspect are often closely correlated in many languages; for example, the English past tense marker -*ed* marks both the past tense and the perfective aspect (Comire 1976). Thus, in our discussion we often need to speak of the acquisition of tense and aspect together.

---

[1] A neural network model is a computational model made of information processing units (neurons) that are connected in a network. The construction and learning of neural network models are often based on considerations of neural information processing. Different from traditional digital computers, the computation in a neural network is based on the connection change among the parallel working units, which has made it a great success in many scientific disciplines during last two decades, such as cognitive science, linguistics, psychology, to name a few. Neural network modeling is also called connectionism or PDP (parallel distributed processing). It views knowledge representation and acquisition as distributed, parallel and interactive in nature. First, a given concept is represented not by a single unit or node but by multiple units or nodes in concert, the result of which is a pattern of activation of relevant micro-features that distribute across multiple units. Second, in terms of knowledge acquisition, connectionism argues for learning through the adaptation of weights, the strengths of connections that hold between multiple and parallel working units, which can also serve as a simplification of the synaptic connections among real neurons. There have been various algorithms developed for adjusting the weights to an optimal set of configurations, which may lead to the appropriate activation patterns of units that represent new knowledge. Third, the interactive activities among multiple units and the learning environment play very important roles in the information processing. For example, PDP argues that linguistic representations can be best understood as the properties that emerge out of learning (i.e., 'emergent properties') rather than built in *a priori*, owing to the interaction of the learning system with the linguistic environment.

1.1. Computational models of tense

The English past tense includes both regular (e.g. *work-worked*) and irregular forms (e.g. *go-went*). When children learn these forms, they sometimes make "over-generalized" error such as producing *goed* as the past tense of *go*, and *breaked* as the past tense of *break*. In addition, studies have shown a "U-shape" trajectory in children's learning of irregular past tenses. At first, they seem to have mastered only a few but correct inflectional forms of verbs; then they forget the correct forms and make many "overgeneralized" errors; finally, children grasp the usage of irregular past tense forms as well as thousands of other regular verbs. (Berko 1958; Elman, Bates, Johnson, Karmiloff-Smith, Parisi, & Plunkett 1996; Marcus, Ullman, Pinker, Hollander, Rosen, & Xu 1992). These phenomena have been traditionally interpreted by some investigators as indicating two totally separate mechanisms for children to learn verb past tenses: one dictionary-like rote mapping for irregular verbs, and the other the explicit representation of a rule of adding suffix *-ed* to regular verbs (Pinker 1994). According to this "dual mechanisms" theory, at first, children learn past tenses only by rote learning, and then almost suddenly, they discover the existence of the rule controlling the formation pattern of regular past tenses; they then apply this rule to any new verb, including irregular verbs, thus causing the so-called "overgeneralized" errors. In the end, children realize that there are exceptions in English past tenses, and then correct their errors and produce the right irregular forms. This type of "dual mechanisms" theory fits well with the general assumption about language as a kind of general symbolic machinery that Chomsky and his followers have advocated.

The symbolic theory of language dominated the psycholinguistic view of morphology and its acquisition for a long time. In 1986, Rumelhart and McClelland (R&M) introduced a simple feed-forward neural network that clearly shows that overgeneralizations and "U-shaped" learning of English past tense may be due to a single mechanism based on neurobiologically plausible features, without the need of two explicit and distinct mechanisms. Basically, the R&M model is a model of pattern associator that can make the strong connection between a verb stem and its phonological form of past tense. There has been much computational modeling work inspired by R&M's model as well as heated debate on what drives the learning of the English past tense.

The R&M model was criticized by Pinker and his colleagues (Pinker & Prince 1988; Marcus etal. 1992) on grounds that the model used unrealistic input and training schedules and the model was unable to capture subtle error patterns in realistic speech. In response to these criticisms, other neural network models equipped with "hidden units" and the "back propagation" learning algorithm have been successfully applied to simulate the acquisition of the English past tense and overcome the shortcomings of the R&M model, such as the model introduced by Plunkett and Marchman (1991) and that by MacWhinney and Leinbach (1991). On the other hand, Ling and Marinov (1993) provided a symbolic pattern associator (SPA) in support of the "dual mechanisms" assumption. The authors claimed that their model outperformed both the R&M model and the MacWhinney & Leinbach (M&L) model when compared with the real data extracted from human subjects. A further detailed comparison of SPA and M&L models (MacWhinney 1993), however, showed that actually the M&L model performed as well as the SPA model in all the past-tense learning tasks; in addition, there were two artificial parameters, which did not have much empirical evidence, that played extremely important roles in the emergence of the U-shaped curve in the SPA model.

In short, in the last two decades, there was a great deal of interest in the computational modeling of the acquisition of tense, focusing on the capacity of neural network models of the English past tense acquisition. The center of the debate was whether the acquisition of grammar can be viewed as the acquisition of symbolic rule systems (in the views of the "dual mechanisms" theory) or whether it can be treated as a statistical learning process (in the views of connectionist theory). The debate is far from being resolved, but the reader is encouraged to consult Elman, Bates, Johnson, Karmiloff-Smith, Parisi & Plunkett (1996) for integrative discussions.

1.2. Computational Models of Aspect

For the remainder of this chapter, we will focus on computational models of the expression and acquisition of aspect, another important temporal concept in languages. Although in the computational linguistics literature, a few computational models have been applied to study aspect categories (as well as tense categories) and analyze the temporal relationships between clauses in terms of event time, speech time, and reference time (Passonneau 1988), computational models of the expression and acquisition of aspect fell far shorter compared with those of the acquisition of tense. Given that previous studies often focused on how to classify verbs into appropriate aspect classes according to the relevant linguistic features and contexts in order to reason about time, we will first discuss here different aspect categorization theories.

1.2.1. Two Kinds of Aspect

Linguists generally distinguish between two kinds of aspect, grammatical aspect and lexical aspect (under various names; see chapter 2, sections 3 and 4, and Li & Shirai 2000, for reviews). Grammatical aspect is related to aspectual distinctions which are often marked explicitly by linguistic devices, such as the inflectional suffixes and auxiliaries in English. It is also known as the viewpoint aspect (Smith 1997) which refers to a particular viewpoint toward the situation being talked about. According to Comrie (1976), there are two major categories of grammatical aspect: imperfective and perfective. Imperfective aspect presents a situation with an internal point of view, often as ongoing (progressive) or enduring (continuous), whereas perfective aspect presents a situation with an external perspective, often as completed. In English, the imperfective-perfective contrast is realized in the difference between the progressive *be V-ing* and the past-perfective *-ed*.

Lexical aspect, on the other hand, refers to the characteristics inherent in the temporal meanings of a verb, for example, whether the verb encodes an inherent end point of a situation, or whether the verb is inherently stative (i.e., continuous and homogeneous) or punctual (i.e., momentary and instantaneous). Most researchers adopt Vendler's (1957) classification as the standard treatment of inherent semantics of verbs, which involves four categories: activities, accomplishments, achievements, and states. A new category has been lately added, which is the so-called "point activities" (Moens & Steedman 1988) or "semelfactives" (Smith 1997). Activity verbs like *walk*, *run* and *swim* encode situations as consisting of successive phases over time with no inherent end point. Accomplishment verbs like *build a house* also characterize situations as having successive phases, but unlike activities they encode an inherent endpoint (e.g., house-building has a terminal point and a result). Like accomplishments, achievement verbs also encode a natural endpoint, but unlike accomplishments and activities they encode events as punctual and instantaneous, that is, as having no duration, such as in *fall*, *recognize a friend* and *cross the border*. State verbs encode situations as homogeneous, with no successive phase or endpoints, involving no dynamicity, such as *know*, *want* and *love*. Finally, the semelfactive verbs involve dynamicity, and encode instantaneous events, but these verbs do not have an inherent end point, like *cough* or *hiccup* in English. In addition, on the basis of whether the verb encodes endpoints, linguists also call activity, state, and semelfactive verbs "atelic" (no endpoint), and accomplishment and achievement verbs "telic" (with endpoint).

In English, grammatical aspect and lexical aspect often interact with each other in complex fashions. Uses of the inflectional suffixes, *-ing, -ed* and *-s* are in many cases constrained. For example, progressive aspect *-ing* does not occur often with state verbs; thus while "John knows the boy" is good, "John is knowing the boy" sounds odd (Smith 1983). There are also combinatorial constraints between *-ing* and event verbs; for example, "The book is falling off the shelf" is odd when used to refer to the actual falling down, but is good when used to mean a preliminary stage (i.e., prior to actual falling; Smith 1997). These kinds of constraints may reflect the intricate relationships between language use and characteristics of the described event. For example, as pointed out by Brown (1973), many events with an end result last for such a short period of time that any description of them is unlikely to occur during the period, such as the actions of *fall, drop,* and *break*. Thus it is rare for speakers to describe the "ongoing-ness" of such events with *-ing* but more natural for them to describe the "completeness" using past-perfective forms.

1.2.2. Computational models of aspectual classification

As we mentioned before, previous computational models about aspect often focused on aspectual classification. For example, in a study conducted by Bennett, Herlick, Hoyt, Liro and Santisteban (1989), using three aspectual features the authors introduced a five-way aspectual classification system to distinguish verbs into the five lexical aspects as we described in section 1.2.1. The three features are ±dynamic, ±atomic and ±telic. Their feature-based descriptions of the five aspectual types are shown as follows (adapted from Bennett et al. 1989):

Accomplishment:	[+d +t −a]
Achievement:	[+d +t +a]
(Extended) Activity:	[+d −t −a]
(Point) Activity/Semelfactive:	[+d −t +a]
State:	[−d −t −a]

This classification method is consistent with Smith's (1991) aspectual analysis except that Bennet et al. used the term "atomic" here to represent the "punctual" feature in Smith's analysis. In addition, the authors further argued that not only the verb features but also some other sentential features (e.g. certain tenses, temporal adverbials) can affect the aspectual situation of verbs in certain sentences. The authors discussed about eleven such sentential feature which can also operate on the [telic] and [atomic] features. For example, in the sentence *John ran*, the verb *ran* keeps its original aspect type of activity. But in some other sentences like *John ran a mile*, J*ohn ran to the park*, and *John ran until 8 o'clock*, the noun phrase, the preposition phrase and the durative adverbial after the verb *ran* imply the endpoint of this action (+telic), and thus the lexical aspect of this verb changes to accomplishment in these sentences.

Bennett et al.'s work showed that aspectual classification is a complex process that depends on both verbs' intrinsic temporal properties and the syntactical features. However, these authors did not show how to use specific methods to realize these classifications computationally. A recent study conducted by Siegel and McKeown (2000) attempted to fill the gap. They declared that the co-occurrence frequencies between the verb and certain linguistic modifiers can reliably predict the verb's aspectual class. The authors generalized 14 so-called "linguistic indicators" (the co-occurrence frequency measures) as the basis to classify verbs into state vs. event verbs, and further into culminated (telic) vs. nonculminated (atelic) verbs in the event category. Specifically, they used three supervised machine/statistical learning methods to combine the 14 linguistic indicators for aspectual classification. The three methods are: logistic regression, decision trees, and genetic algorithm (GA).

Logistic regression is a multivariate statistical method that can derive an overall variate based on the weighted nonlinear combination of the 14 variables or the linguistic indicators. The overall variate can increase the classification performance of the model. The decision tree is a traditional data mining method based on many choice points. At each point, according to the value of a specific linguistic indicator, the system makes an *if-then-else* choice to decide which one of the two possible classes a verb should belong to. When facing a classification task for a verb, the method will start from the root of the decision tree, undergo a series of tests on choice points, and then end at a leaf (thus labelled by an aspectual type). The decision tree enables the complex interaction of different indicators in the system. Finally, genetic algorithm (GA) is a novel method in computer science based on the concept of Darwinian natural selection and survival of the fittest (Holland 1975). This method enables the generation and evolution of arbitrary mathematical combinations of the 14 linguistic indicators to classify the verb aspects. All the three methods have shown good performance on the two-way classification of lexical aspect (state vs. event; culminated vs. nonculminated).

Neural network models can also be used in aspectual classifications. Based on the idea that people should be able to extract aspectual features and meanings from syntactic representations, Scheler (1997) introduced a model that can extract the grammatical categories of English aspects (progressive vs. simple) and Russian aspects (imperfective vs. perfective). The model includes four main modules:

(1) An automatic tagger that can provide syntactic tags to the words in the input text in the system, thus transferring the text to specialized syntactic representations;
(2) A process that transforms the syntactic representation into the semantic representation, a syntactic-to-semantic pattern association task;
(3) A set of semantic features describing aspectual meanings of verbs, such as event type, action status, and habituality;
(4) A process that maps the individual aspectual meanings to the grammatical categories for each language, which is the final pattern classification task.

The modules (2) and (4) are the core parts of the model, and Scheler used two standard back-propagation neural networks with hidden layers to simulate the processes of the two modules. For module (2), the author constructed a network with an input layer of 25 binary neurons (which represent the 6 slots syntactic features), two hidden layers with 15 and 12 neurons respectively, and an output layer with 34 binary neurons representing the 15 semantic features. The network was trained to associate the syntactic patterns of verbs to their semantic representations. The network successfully learned 87 percents of the total syntactic-semantic pattern pairs, although the generalization performance was not as good as the learning. Based on the simulating results, the author argued that most of the information needed to extract semantic features for aspects is based on local syntactic features. For module (4), the author used a $34 \times 5 \times 2$ network to classify verbs into different grammatical aspectual categories according to their semantic features. The network has an input layer with 34 binary neurons, a hidden layer with 5 neurons, and an output layer with 2 neurons to represent the grammatical aspectual types. The network performed well in both learning and generalization of the patterns. This model was a first full-scale back-propagation connectionist model for aspectual classification. A problem with the model is that the two neural networks for modules (2) and (4) in the model were isolated and did not communicate with each other. In addition, the generalization ability of the syntactic-semantic association network was not good. In section 2, we will introduce a self-organizing neural network model that can correct these problems.

The aspectual models discussed above are all about the expression or classification of aspectual categories. In contrast to this line of research, scholars have also been interested in how children acquire the classification of lexical aspect and how this acquisition influences their use of grammatical aspect, which brings us to the next section.

1.2.3. Aspect acquisition in child language

A core issue in the study of aspect acquisition is how children acquire the two kinds of aspect (grammatical aspect and lexical aspect, see section 1.2.1) and their interactions in different languages. It has been now well established that there is a strong association between lexical aspect and grammatical aspect in child language: children initially tend to restrict tense-aspect morphology to specific categories of lexical aspect. For example, English-speaking children initially tend to use progressive marker *-ing* only with atelic, activity verbs, whereas past-perfective marker *-ed* only with telic verbs (accomplishment and achievements) at an early stage of development (McShane & Whittaker 1988; Shirai & Andersen 1995). This restricted or "undergeneralized" pattern of use has led to intense debate with respect to various theoretical frameworks (see Li & Shirai 2000 for review). An early suggestion from Bickerton (1984) was that children have innate semantic categories that roughly correspond to the lexical aspect distinctions of verbs (e.g., punctual-nonpunctual, state-process distinctions), and these categories are biologically programmed as part of a Language Bioprogram. Bickerton relied on both data from creole languages and child language acquisition to support his proposal that children's acquisition of tense-aspect morphology has a biological basis. Subsequent crosslinguistic studies, however, have provided counter evidence to this hypothesis (e.g., Li & Bowerman 1998; Shirai & Andersen 1995), and led researchers to propose a variety of input-driven hypotheses about how children acquire tense-aspect morphology and lexical semantics of verbs (see Li & Shirai 2000 for a review).

The goal of our computational modelling is to provide mechanistic accounts of how empirically observed patterns could emerge out of simple computational principles. In previous empirical studies (Li & Bowerman 1998; Li & Shirai 2000), we proposed that the initial lexical-morphological associations could arise as a result of the learner's analyses of the verb-morphology co-occurrence probabilities in the input environment of the language learner. In parental speech, there are probabilistic associations between progressive markers and atelic verbs, and between perfective markers and telic verbs (Shirai & Andersen 1995). Children's initial undergeneralizations (restricted uses of morphology) might reflect their analyses of these probabilities. In the next section, we will discuss a neural network model which can analyze these probabilities and arrive at patterns that resemble children's patterns of acquisition.

2. A self-organizing neural network model of the acquisition of aspect

In the last few years we have explored self-organizing neural networks as candidates of cognitively and neurally plausible models of language acquisition (Li 2003, 2006; Li, Farkas & MacWhinney 2004; Li, Zhao & MacWhinney 2007; Zhao & Li 2005, 2007, 2008, in press). Compared to other developmental neural network models, most of which rely on supervised learning algorithms (e.g., back-propagation, see models reviewed in Elman et al. 1996), self-organizing neural networks, especially the so-called self-organizing maps (SOM), have several important properties that make them particularly well suited to the study of lexical and morphological acquisition (see Li 2003, 2006 for discussion).

First, these models belong to the class of unsupervised learning networks that require no explicit teacher; learning is achieved by the system's organization in response to the input. Such networks provide computationally more relevant models for language acquisition, given that in real language learning children do not receive constant feedback about what is incorrect in their speech (see Li 2003; MacWhinney 1998, 2001; Shultz 2003 for discussion). Second, self-organization in these networks allow for the gradual formation of structures on 2-D maps, as a result of extracting an efficient representation of the complex statistical regularities inherent in the high-dimension input space (Kohonen 2001). Third, the self-organizing map forms topography-preserving structures, which means nearby areas in the map respond to inputs with similar features. This property allows us to model the emergence of semantic categories as a gradual process of lexical learning. Finally, several self-organizing maps can be connected via Hebbian learning, a well-established biologically plausible learning principle, according to which the association strength between two neurons is increased if the neurons are both active at the same time (Hebb 1949). Although Hebbian learning itself is not an inherent property of the self-organizing algorithm, when incorporated, the SOM model would have strong implications for language acquisition: it can account for the process of how the learner establishes relationships between word forms, lexical semantics, and grammatical morphology, on the basis of how often they co-occur and how strongly they are co-activated in the representation.

A number of recent models have taken advantage of the properties discussed above to examine language processing and language acquisition. These include DISLEX (Miikkulainen 1997), DevLex (Li, Farkas, & MacWhinney 2004), DevLex-II (Li, Zhao & MacWhinney 2007), and SEMANT (Silberman, Bentin, & Miikkulainen 2007). In particular, we have applied the DISLEX and DevLex-II models to the study of the acquisition of grammatical aspect (*-ing*, *-s* and *-ed*) in connection with the acquisition of semantic categories of lexical aspect (Li 2000; Li & Shirai 2000; Zhao & Li, in press). In what follows, we will provide a review of the findings from our DevLex-II model that has been used to simulate aspect acquisition[2] (Zhao & Li, in press).

2.1. A sketch of DevLex-II

DevLex-II is a multi-layer self-organizing neural network for modeling early lexical acquisition. It is based on and adapted from the DevLex model (Li, Farkas, & MacWhinney 2004). It has been developed to account for empirical phenomena in early lexical acquisition (e.g., 'vocabulary spurt') and bilingual lexical development (Li, Zhao, & MacWhinney 2007; Zhao & Li 2005, 2007, 2008). The basic structure of DevLex-II is shown in Figure 1.

---

[2] Here we only provide a review of simulation results based on DevLex-II. For detailed results about aspect acquisition with the DISLEX model, see Li (2000) and Li and Shirai (2000: ch. 7).

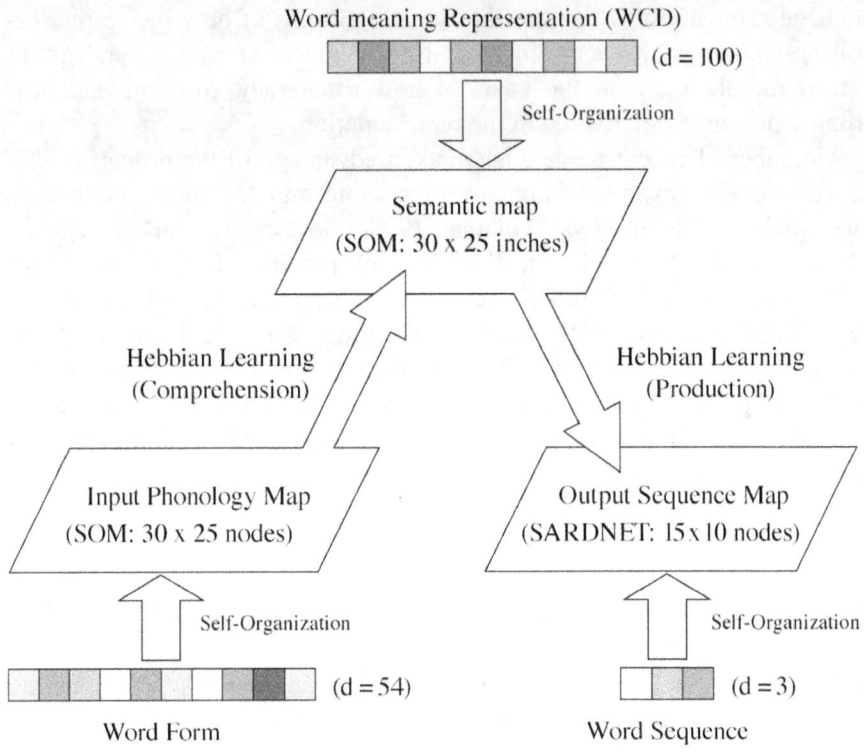

Figure 1. The architecture of the DevLex-II model. Each of the three self-organizing maps (SOM) takes input from the lexicon and organizes phonology, semantics, and phonemic sequence information of the vocabulary, respectively. The number of nodes in each map is indicated in parentheses. The dimension of the input vector for each map is indicated by 'd = ' in parentheses next to the input representation symbols. The maps are connected via associative links updated by Hebbian learning.

DevLex-II uses three layers of SOMs to process three basic levels of linguistic information: phonological content, semantic content, and output phonemic sequence. The phonological layer and semantic layer operate according to the standard SOM algorithm (see Kohonen 2001, for details). The standard SOM constructs a two-dimensional topographic map for the organization of input representations, where each node (or "neuron") is a location on the map that has input connections to receive external stimulus patterns. At each training step of SOM, an external input pattern (e.g., the phonological or semantic information of a word in our study) is randomly chosen and presented to all the nodes on the map; this activates many nodes on the map, according to how similar by chance the input pattern is to the weight vectors of the nodes, and the node that has the highest activation is declared the winner (the Best Matching Unit or BMU). Once a node becomes active in response to a given input, the weight vectors of that node and its neighboring nodes are adjusted, so that they become more similar to the input and the nodes will respond to the same or similar inputs more strongly the next time. In this way, every time an input is presented, an area of nodes will become activated on the map (the "activity bubbles") and the maximally active nodes are taken to represent the input. Initially activation occurs in large areas of the map, that is, large neighborhoods, but gradually learning becomes focused and the size of the neighborhoods reduces. This process continues until all the inputs have found some maximally responding nodes as their BMUs. As a result of this self-organizing process, the statistical structures implicit in the input are represented as topographical structures on the 2-D space. In this new representation, similar inputs will end up activating nodes in nearby regions, yielding meaningful activity bubbles that can be visualized on the map.

The addition of the phonemic sequence layer represents a step forward from the original DevLex model, and is inspired by models of word learning based on temporal sequence acquisition (e.g., Gupta & MacWhinney 1997). It is designed to simulate the challenge to young children when they need to develop better articulatory control of the phonemic sequences of words. Just as the learning of auditory sequences requires the mediation of memory systems, the learning of articulatory sequences requires support from the rehearsal in phonological working memory (Gathercole & Baddeley 1993; Gupta & MacWhinney 1997). In our implementation of this idea, the activation pattern corresponding to the phonemic sequence information of a word is formed according to the algorithm of SARDNET (James & Miikkulainen 1995), which works slightly differently from the standard SOM algorithm. At each training step, phonemes are input into the sequence map one by one, according their order of occurrence in the word. The winning unit of a phoneme is found and the responses of nodes in its neighborhood are adjusted. Once a unit is designated as the winner, it becomes ineligible to respond to the subsequent inputs in the sequence. In this way, same phonemes in different locations of a word will be mapped to different (but adjacent) nodes on the map as a result of the network's topography-preserving ability. When the output status of the current winner and its neighbors is adjusted, the activation levels of the winners responding to phonemes before the current phoneme will be adjusted by a number γ^d, where γ is a constant and d

is the distance between the locations of the current phoneme and the previous phoneme that occurred in the word. This adjustment is intended to model the effect of phonological short-term memory during the learning of articulatory sequences; the activation of the current phoneme could be accompanied by some rehearsal of previous phonemes due to phonological memory, which deepens the network's or the learner's impression of previous phonemes. The γ here is chosen to be less than 1 (0.8 in our case), in order to model the fact that phonological memory tends to decay with time. For further details of the DevLex-II model, see Li, Zhao, and MacWhinney (2007).

The associative links between any two layers of maps in DevLex-II are trained by Hebbian learning, such that the activation of a word on the form map can evoke the activation of a word on the meaning map via form-to-meaning links, thereby modeling word comprehension, and the activation of word meaning can cause the formation of word sequence via meaning-to-sequence links, thereby modeling word production. In DevLex-II, we say that a word has been learned in comprehension when a node in the destination map (word meaning map) becomes consistently activated as the 'winner' for a given input from the source map (word form map). We say that a word has been learned in production when several nodes in the word sequence map become activated sequentially as winners that represent the word's phonemes.

2.2. Input representations for DevLex-II

As with DevLex, we used the PatPho system to construct the phonological patterns for word forms. PatPho is a generic phonological pattern generator for neural networks, which fits every word (up to tri-syllables) onto a template according to its vowel-consonant structure (Li & MacWhinney 2002). PatPho uses the same phonological method as in MacWhinney and Leinbach (1991), but relies on articulatory features of phonemes (Ladefoged 1982) to represent the phonemes, *C*s and *V*s, and a phoneme-to-feature conversion process to produce real-value or binary feature vectors for any word up to three syllables. In short, PatPho can code each input word in our simulation by the template *CCCVVCCCVVCCCVVCCC*, and then replace each phoneme with its appropriate representation using real-value or binary numbers.[3] For example, the verb *pick* with its progressive marker *-ing* would be

---

[3] In this simulation we used the real-value vectors.

encoded as p*CC*ɪ*Vk*C*C*ɪVŋ*CCVVCCC*, and is represented as the following vector:

/pɪkɪŋ/: 1–0.45–0.733   0–0–0   0–0–0   0.1–0.1–0.185   0–0–0   1–0.921–0.733   0–0–0
          0–0–0   0.1–0.1–0.185   0–0–0   0.75–0.921–0.644   0–0–0   0–0–0   0–0–0   0–0–0
          0–0–0   0–0–0   0–0–0

In this representation, the first three units [1–0.45–0.733] indicate the phonetic features of phoneme /p/, the second and third sets of three units indicate that no more consonants follow /p/ in this word (hence zeros). The representation is left-justified, which means that in a given syllable, the representation of the phoneme is pushed toward the left side of the template (rather than the right side).

On the output sequence map, the phonemes of a word are processed one by one, so we need representations for each of the 38 English phonemes. Using the method of PatPho, we can represent these phonemes by three-dimensional real-value vectors. In particular, in the vector, the first dimension indicates whether the phoneme is a vowel or a consonant, and in the case of a consonant, whether it is voiced or voiceless. The second dimension indicates the position for vowels and manner of articulation for consonants and the third dimension indicates the sonority for vowels and place of articulation for consonants (see Li & MacWhinney 2002).

With respect to the semantic representation of the input, we used a special recurrent network called WCD (word co-occurrence detector) to generate vectors. WCD allows us to generate vectors that dynamically change with the learning history: lexical representations enrich over time as a function of learning the number of co-occurring words in the input sentences. Metaphorically, this learning scenario can be compared to filling the holes in a Swiss cheese: initially there may be more holes than cheese (*shallow* representations) but the holes get filled up quickly as the co-occurrence context expands with more words being acquired (*rich* representations).

Briefly, WCD works as follows (see Farkas & Li 2001, 2002; Li, Farkas & MacWhinney 2004, for details). It reads through a stream of input sentences one word at a time, and learns the transitional probabilities between words which it represents as a matrix of weights. Given a total lexicon sized N, all word co-occurrences can be represented by an $N \times N$ contingency table, where the representation for the i_{th} word is formed by concatenation of i_{th} column vector and i_{th} row vector in the table. Hence, the two vectors correspond to the left and the right context, respectively; WCD transforms these probabilities into normalized vector representations for word meanings, which in turn are read by self-organizing maps (after Random Mapping, a procedure to achieve reduced uniform dimension of vectors; see Ritter & Kohonen 1989). Here, different inflectional forms of the same verb are considered as different items, and therefore WCD will derive different (but also similar) representations for them. For example, *playing* and *played* will be represented distinctly, but since the co-occurrence contexts for these words will overlap significantly (e.g., the co-occurring words tend to be *ball, toys,* etc.), the representations for them will also tend to be similar. An example of a semantic representation generated by WCD is shown below (only part of the vector is shown, the full vector contains 2002 units). Here, every two units of the vector represent the normalized co-occurrence possibility between the verb *picking* and another word in the lexicon. The odd unit represents the possibility that *picking* happens before a given word, and the even unit represents the possibility that *picking* follows a given word in the context.

Picking: *0.000000   0.000000   0.006004   0.007211   0.003548   0.017577   0.000000*
0.000000   0.000000   0.000000   0.000000   0.000000   0.000000   0.001202
0.000000   0.000300   0.000000   0.000000   0.000000   0.000000 ...

DevLex-II also uses as its input data the parental or caregivers' speech in the CHILDES database (MacWhinney 2000). Here we extracted all parental or caregivers' utterances from the complete English database (as of 2002). A verb type was chosen as input if it occurred in the parental speech for fifty or more times in a given age period. The verbs were divided into four stages to be presented to the network, according to the age groups (Age 1;6, 2;0, 2;6, 3;0) at which they occurred. To increase the accuracy of WCD representations, we also analyzed the selected verbs along with the nouns, adjectives, and closed-class words from the MacArthur-Bates Communicative Development Inventories (the CDI, Toddler's List; Dale & Fenson 1996; homographs and homophones, word phrases, and onomatopoeias were excluded). These CDI words along with the verbs that fit our selection criterion (a total of 1001 words) served as the input contexts of WCD. We computed the semantic representations of the vocabulary at each of the four growth stages, resulting in 4 different data sets with increasing complexity in semantic representation. The four growth stages had the following vocabulary composition:

(1) Input Age 1;6 (13–18 months): a total of 62 verb types fit our selection criteria for the period before age 1;6; 35 of these verbs occurred with *-ing*, 13 with *-ed*, 14 with *-s*.

(2) Input Age 2;0 (19–24 months): 100 verb types were selected, which included the new words as well as words from the previous stage; 58 occurred with -*ing*, 19 with -*ed*, 23 with -*s*.
(3) Input Age 2;6 (25–30 months): 154 verb types were selected, among which 86 occurred with -*ing*, 32 with -*ed*, 36 with -*s*.
(4) Input Age 3 (31–36 months): A total of 184 verb types were selected, out of which 97 verbs occurred with -*ing*, 41 with -*ed*, and 46 with -*s*. This stage included all verbs that occurred in previous stages plus new ones.[4]

As shown in Figure 1, the size of the network was 30 x 25 nodes for the phonological map and the semantic map, and 15 x 10 nodes for the phonemic map. These numbers were chosen to be large enough to discriminate among the words and phonemes in the lexicon, while keeping the computation of the network tractable. For each training stage, the network was trained for 50 epochs, which means that each verb in a given stage was presented to each map 50 times.

2.3. Association of lexical and grammatical aspect in DevLex-II

An important rationale behind our simulations is for us to understand the role of linguistic input in guiding children's acquisition of lexical and grammatical aspect. Here, we wanted to test whether DevLex-II, endowed with self-organization and Hebbian learning principles, is able to display learning patterns as the child does. Our networks receive phonological and semantic representations of input words based on actual adult speech along with phonemic sequence (morphology) information of these words. If the network is able to produce patterns like those we found in children's speech on the basis of learning of the input, we can then conclude that self-organization and Hebbian learning provide the necessary kinds of mechanisms that drive the formation of patterns in children's acquisition. In this way, our modeling enterprise sheds light on the mechanisms that underlie the learning process.

---

[4] Bare verb forms were excluded from our simulations, as well as irregular past tense forms and non-verbs. Exclusion of these forms simplifies the simulation task and makes the analysis more tractable. Our major goal here is to demonstrate whether the use of verbal suffixes is correlated with the lexical aspect of verbs, and as such our simulations focused on suffixed verbs.

Table 1 (adapted from Table 2 of Zhao & Li, in press) provides a summary of the major patterns from the DevLex-II models, according to the tense-aspect suffixes the model produced at different learning stages. The table presents the results of the networks' production of three suffixes, *-ing, -ed,* and *-s,* with three types of verbs, activity, telic and stative.[5] The results were based on the analysis of the networks' production ability; that is, how semantic representations induce activations on corresponding feature map (phonemic sequence map in DevLex-II) through associative pathways. The analysis was done by inspecting the nodes that each verb on the semantic map activated, after the network had been trained for a specified number of epochs at each stage (50 epochs for DevLex-II).

The testing of DevLex-II's word production ability is as follows. At the end of each training stage, verb types in the lexicon of the current stage are presented to the semantic map one by one. For a verb type, its best matching unit or BMU on the semantic map is found, and in turn this node propagates its activation to the output sequence map through the associative links. Several nodes in the sequence map become activated sequentially as winners that represent the word's phonemes. Then the network checks to see if every node is the BMU of a unique phoneme, according to the Euclidean distance between its input weight vector and the feature representation of every phoneme. If it is, the phoneme closest in Euclidean distance to the current winner becomes its retrieved phoneme; if it is not, the pronunciation of this phoneme has failed. Finally, the pattern of the retrieved phoneme sequence is treated as the output of word production. When the retrieved phonemic sequence matches up with the actual word's phonemic sequence, we say that the word has been correctly produced. For example, if the word *kicking* is shown to the semantic map, correct production occurs only when the consecutively activated nodes on the output phonemic map are the BMUs for /k/ /ɪ/ /k/ /ɪ/ /ŋ/ in this particular sequence.

---

[5] Our analyses below deviate slightly from a strict five-way classification, because accomplishment and achievement verbs are often difficult to separate without an extensive analysis of the sentence and speech context. Thus in what follows telic verbs include both accomplishments and achievements.

Table 1. Percentage of use of tense-aspect suffixes with different verb types across input age groups in DevLex-II's production and in parental input data (including verbs with multiple suffixes)

	TENSE-ASPECT SUFFIXES											
	Age 1;6			Age 2;0			Age 2;6			Age 3;0		
VERBS	-ing	-ed	-s	-ing	-ed	-s	-ing	-ed	-s	-ing	-ed	-s
Network Production												
Activity	73	0	29	69	27	33	61	24	35	62	30	37
Telic	27	75	14	21	53	28	32	62	27	31	62	26
Stative	0	25	57	10	20	39	7	14	38	7	8	37
Parental Input Data												
Activity	63	23	29	62	26	26	63	22	33	60	29	35
Telic	31	62	29	31	58	26	29	66	25	32	59	24
Stative	6	15	43	7	16	48	8	12	42	8	12	41

The results of Table 1 are highly consistent with empirical patterns observed in early child language: the use of imperfective aspect is closely associated with activity verbs that indicate ongoing processes, while the use of perfective aspect is closely associated with telic verbs that indicate actions with endpoints or end results. In particular, in early child English, the progressive marker *-ing* is highly restricted to activity verbs, the perfective/past marker *-ed* restricted to telic verbs, and the third person singular *-s* restricted to stative verbs (Bloom, Lifter and Hafitz 1980; Brown 1973; Clark 1996; Shirai 1991). Our network, having taken in input patterns based on realistic parental speech, behaved in the same way as children do. For example, at Input Age 1;6, the networks produced *-ing* predominantly with activity verbs (73%), *-ed* overwhelmingly with telic verbs (75%), and *-s* with stative verbs (57%). Such associations were strong at all four stages (especially for *-ing* and *-ed*), but they tended to become weaker over time.

Interestingly, when we analyzed the actual input to our networks (based on parental speech), we found similar patterns. Table 1 also shows the percentages of the use of suffixes with different verb types in the input data for DevLex-II. An analysis of the table indicates that in the input data there are also clear associations between *-ing* and activity verbs, *-ed* and telic verbs, and that these associations are strong throughout the four stages, as also found previously by Shirai (1991) and Olsen, Weinberg, Lilly and Drury (1998). The degree to which the networks' production matches up with the input patterns indicates that DevLex-II was able to learn on the basis of the information of the co-occurrences between lexical aspect (verb types) and grammatical aspect (verb morphology). This learning ability was due to the networks' use of Hebbian associative learning in computing if the semantic, phonological, and phonemic properties of a verb co-occur and how often they do so.

To see the data more clearly, we illustrate the patterns with Figure 2[6] to show the percentages of the use of suffixes with different verb categories in both DevLex-II's productions (Figure 2a) and in parental input data (Figure 2b) at Input Age 1;6. Comparing Figure 2a and 2b, we can see that the network's production patterns are consistent with patterns in the parental input data, but the network showed more restricted use of the suffixes rather than a verbatim replication of the input association patterns.

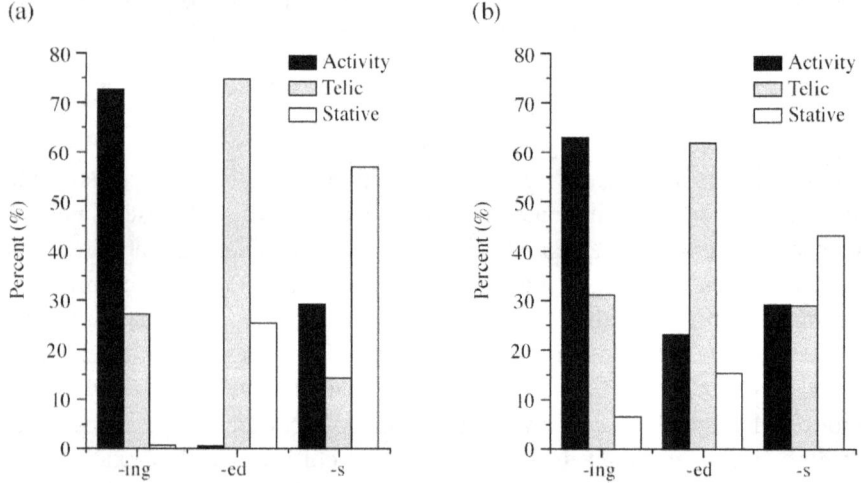

Figure 2. Percentages of the use of suffixes with different verb types at Input age 1;6 in (a) network productions in DevLex-II, and (b) parental input data. Data are based on Table 1.

2.4. Structured semantic representations of aspect in DevLex-II

Elsewhere we have proposed an account of semantic development as an emergent process in which semantic features are connected in a system to support lexical categories, like in the formation of semantic cryptotypes (Li & MacWhinney 1996; Li 2003; Li, Farkas & MacWhinney 2004; Hernandez,

---

[6] Figures 2–4 in this chapter were adapted from Zhao & Li (in press).

Li, & MacWhinney 2005; see also Rogers & McClelland, 2004, for similar discussions). The basic idea is that a given verb may be represented with multiple linguistic features, and the features themselves often co-occur and overlap in different verbs. For example, the verb *screw* may be viewed as having both a meaning of circular movement and a meaning of binding or locking, and the verb *zip* may be viewed as sharing both the "binding/locking" meaning and the "covering" meaning. Moreover, both *screw* and *zip* involve hand movements. Features may also vary in the strength with which they are represented in different verbs. For example, the verb *wrap* may be viewed as having the covering meaning. However, in some cases, the action of wrapping may also involve circular movements. Children may acquire such complex feature-to-verb relationships through statistical analyses of the co-occurrences of verbs with situational contexts, with other words, and co-occurrences of particular grammatical morphemes with semantic features (see discussion in Siegel and McKeown's 2000 work reviewed in section 1.2.2), leading to feature-based organization of verb categories. In the simulations here, we provided our networks with verbs that are represented with multiple semantic/syntactic features (lexical co-occurrence constraints, extracted by WCD), and we wanted to see how categories of lexical aspect could emerge from the self-organizing learning process.

A particularly useful property of self-organizing feature maps is that the statistical structures in the representations can be clearly visualized as activity bubbles or patterns of activity on a two dimensional map in a topography-preserving structure. Given that DevLex-II represented semantic information from the high-dimensional space of verb usage in parental input, we hypothesize that verbs with similar aspectual properties should cluster together on the feature map. Figure 3 present a snapshot of DevLex-II's self-organization of the semantic representations of verbs (with suffixes) at the end of the learning process (i.e., Stage 4, Input Age 3;0).

Figure 3. Emergent semantic representations in DevLex-II after Input age 3;0. Differently shaded regions indicate different aspect categories corresponding to different suffixes -*ed*, -*ing*, and -*s*. Within each category, verbs with the same lexical aspect are often grouped together; see text for discussion.

An examination of this map shows that the network has clearly developed structured semantic representations that correspond to different lexical aspect categories. It formed clear clusters of verbs by mapping verbs with similar combination of semantic features onto nearby regions of the map. We can make several interesting observations on the basis of these results:

(1) The most obvious structure of the map is that the words can be roughly divided into three main clusters according to the suffix that a verb stem takes, -*ing*, -*ed*, or -*s* (see Figure 3);
(2) Within each cluster, there are also groups that correspond to categories of lexical aspect such as telic verbs, activity verbs, and stative verbs. For example, towards the lower left-hand corner of the larger part of the -*s* cluster (the light gray area), stative verbs, like *loves, knows, likes,*

wants, and *needs* are mapped to the same region. Another example can be found in the *-ing* cluster (the area without shading): although most verbs clustered in this area are activity verbs such as *working, sitting, crawling, walking, sleeping, etc.*, there is also a cluster of telic verbs (at the middle-to-lower portion of the map) such as *wiping, fixing, hitting, putting, cutting, throwing, making,* and *getting*;

(3) The distribution of lexical aspect is closely related to the distribution of grammatical aspect. Not only it is the case that the *-ed* cluster contained mostly telic verbs and the *-ing* cluster mostly activity verbs, but also telic verbs that take *-ing* were closer to the *-ed* cluster (e.g., *going, jumping, messing, picking* and *cleaning*, all bordering the *-ed* cluster);

(4) Verbs with the same stem but different suffixes are also often mapped to regions not far away from one another, for example, *fixing* and *fixed*, *pushing* and *pushed*, *turns* and *turned* at the middle area of the map, and *playing* and *played* at the lower right corner of the map.

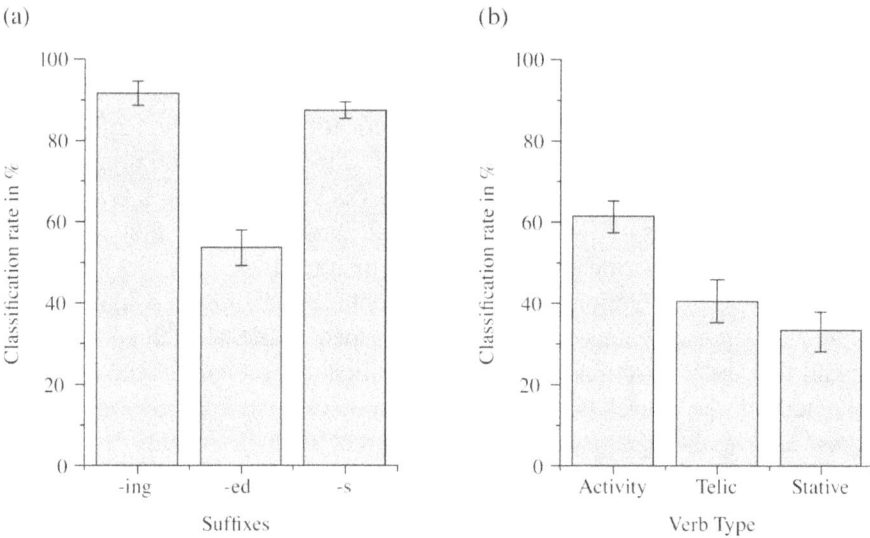

Figure 4. Classification rates calculated by a 5-NN classifier according to lexical and grammatical representations of the verbs in DevLex-II's semantic map. Classifications are based on: (a) the suffix that a verb stem takes: *-ing*, *-ed*, or *-s*; (b) the lexical aspect of the verb: activity, telic, or stative. The error bars indicate the standard deviations based on 5 trials.

The emergence of structured semantic representations in our model can also be verified by a simple method called k-nearest neighbor (k-NN) algorithm (Duda, Hart & Stork 2000). As a classical method in the field of pattern recognition for classifying objects into different classes, the basic idea of k-NN is to predict the class of a point in a dataset according to the most frequent class label of its k nearest neighbors. Implementing this method in our semantic map (see also Li, Farkas & MacWhinney 2004), we can evaluate if a verb in our lexicon was mapped to a node close in Euclidean distance to other verbs belonging to the same class. This allows us to have a rough idea of the overall compactness of different lexical classes. Here, we conducted a 5-NN analysis of verb representations on the semantic map according to the suffix a verb stem takes, *-ing, -ed,* or *-s*. As shown in Figure 4a, the semantic map has developed clear clusters for different suffixes: for the category of *-ing*, the classification rate is about 92%, which means that 92 percents of verbs suffixed with *-ing* are located within a nearest neighborhood according to k-NN; for the categories of *-s* and *-ed*, the classification rates are 88% and 60%, respectively. We also conducted a 5-NN analysis of the verbs according to their lexical aspect properties: as shown in Figure 4b, the classification rates for activity verbs, telic verbs, and stative verbs are 61%, 41%, and 34%, respectively. The relatively low classification rates of the verb categories, compared with those of suffix categories, indicate that the organization of verb meanings according to lexical aspect is subordinate to the organization of verb suffixes on the map. In general, these quantitative analyses are consistent with our visual analyses of the semantic maps.

The above observations lead us to conclude that the map has formed structured representations for grammatical aspect markers, such as *-ed, -ing, -s*, and that the interaction between grammatical aspect and lexical aspect is reflected in the correlation between grammatical morphology and verb types, and in the categories of lexical aspect such as activity, telic, and stative. The results from our modeling offer a new way of thinking about the representation of lexical aspect and its interaction with grammatical aspect. Verbs in a lexical aspect category form complex relationships, in that they vary in (1) how many linguistic features are relevant to the category, (2) how strongly each feature is activated in the representation of the category, and (3) how features overlap with each other across category members. For example, *spill* may be viewed as indicating both a punctual and a resultative meaning; *close* may involve both a change of state and a completive meaning; and the feature "punctual" may be represented more strongly in *jump* than in *fall*: in a natural setting a single jump occurs instantaneously, whereas falling need not (e.g., we could still say that a leaf fell from a tree even if it drifted down slowly).[7] With varying degrees of connections from semantic features to verb forms, verbs can form clusters or categories that differ overall in lexical aspect. Traditional analytical methods from linguistics and psycholinguistics are much less effective in dealing with these complex semantic relationships (but see Siegel and McKeown's recent attempt to combine different linguistic indicators, reviewed earlier). By contrast, neural network models that rely on distributed feature representations and nonlinear learning are ideally suited to accounting for the properties of feature overlapping and weighted feature composition. Devlex-II provides a clear example for how we may solve complex semantic problems via weighted feature composition (see also Li & MacWhinney 1996; Li 2003 2006; Li, Farkas & MacWhinney 2004).

3. Conclusion

In this chapter, we presented an overview of the computational models of the expression and acquisition of temporality in languages. For the issue of tense, we provided a brief introduction to computational modeling of children's acquisition of the English past tense, and the heated debate on single versus dual mechanisms revolving around this issue. For the issue of aspect, we reviewed a few computational models of aspectual classification, and introduced in detail our DevLex-II model for simulating aspect acquisition in languages. Our review clearly shows that the expression and acquisition of aspect are very complex processes that depend both on the verb's internal semantic meanings about temporal concept and on the syntactic features of the sentence where the verb occurs. Our DevLex-II model successfully simulates the acquisition of lexical and grammatical aspect, and provides insights into issues regarding the role of linguistic input, the emergence of lexical categories of verbs, and the development of prototypical to non-prototypical associations.

Self-organization and Hebbian learning in our model are two important computational principles that can account for the psycholinguistic processes in the acquisition of lexical and grammatical aspect. Our simulations dem-

---

[7] To see this, in Figure 3, we can find that the word *jump*, with its progressive marker *-ing*, is much closer than *falling* to the *-ed* cluster. Note that *jump* can also be construed iteratively, so that *jumping* refers to a series of jumps, which is why we may see that children use *jumping* more frequently than *jumped* in natural speech.

onstrate that the network is able to display patterns of association as observed in empirical acquisition studies, on the basis of its analyses of input characteristics. In particular, self-organization of the semantic structure of verbs leads to the formation of lexical aspect categories and grammatical aspect categories, on the basis of the network's analysis of the complex feature-to-verb and verb-to-morphology relationships in language use. In addition, our model clearly shows that simple but biologically plausible computational principles in self-organizing neural networks can account for empirically observed patterns in children's acquisition of lexical aspect and grammatical morphology, without a priori stipulations about the structure of meaning or concept.

Contributing to the debate on single versus dual mechanisms for learning, our model also specifically suggests that the learning of grammatical suffixes is not simply the learning of a rule (such as adding -*ing* or -*ed* to a verb to mark the progressive aspect or the perfective aspect), but the accumulation of associative strengths that hold between a particular suffix and a complex set of semantic features distributed across verb forms (which support the emergence of a lexical aspect category). This learning process can be best described as a statistical, probabilistic process in which the learner implicitly tallies and registers the frequency of co-occurrences (strengthening what goes with what) and co-occurrence constraints (inhibiting what does not go with what) among the semantic features, lexical forms, and tense-aspect suffixes. The co-occurrence-and-constraint process is clearly modeled in our network by Hebbian learning of the associative connections between forms and meanings (see also a detailed description in Zhao & Li, in press).

To conclude, the models we reviewed here clearly serve to demonstrate the utility of computational modeling (especially connectionist modeling) for unraveling mechanisms underlying the expression and acquisition of tense and aspect in languages. With the rapid development of computing techniques and the advancement of computational modeling, we are hopeful that the detailed cognitive and psycholinguistic mechanisms can be clearly revealed in the fields of language acquisition, language representation, and language processing.

References

Bates, Elizabeth, Jeffrey Elman, and Ping Li
    1994   Language in, on, and about time. In *The Development of Future-Oriented Processes*, M. Haith, J. Benson, R. Roberts and B. Pennington (eds.), 293–321. Chicago, IL: University of Chicago Press.

Bennett, Winfield, Tanya Herlick, Katherine Hoyt, Joseph Liro and Ana Santisteban
    1989   Toward a computational model of aspect and verb semantics. *Machine Translation* 4: 247–280.

Berko, Jean
    1958   The child's learning of English morphology. *Word* 14: 150–177.

Bickerton, Derek
    1984   The language bioprogram hypothesis. *Behavioral and Brain Sciences* 7: 173–188.

Bloom, Lois, Karin Lifter, and Jeremie Hafitz
    1980   Semantics of verbs and the development of verb inflection in child language. *Language* 56: 386–412.

Brown, Roger
    1973   *A First Language.* Cambridge, MA: Harvard University Press.

Clark, Eve V.
    1996   Early verbs, event-types, and inflections. In *Children's Language*, Vol. 9, C. E. Johnson & J. H. V. Gilbert (eds.). Mahwah, NJ: Erlbaum.

Comrie, Bernard
    1976   *Aspect: An Introduction to the Study of Verbal Aspect and Related Problems.* Cambridge: Cambridge University Press.

Dale, Philip S. and Larry Fenson
    1996   Lexical development norms for young children. *Behavior Research Methods, Instruments, & Computers* 28: 125–127.

Duda, Richard O., Peter E. Hart, and David G. Stork
    2000   *Pattern Classification*, 2nd Ed.. New York, NY: John Wiley & Sons.

Elman, Jeffrey L., Elizabeth A. Bates, Mark H. Johnson, Aannette Karmiloff-Smith, Domenico Parisi, and Kim Plunkett
    1996   *Rethinking Innateness: A Connectionist Perspective on Development.* Cambridge, MA: MIT Press.

Farkas, Igor and Ping Li
    2001   A self-organizing neural network model of the acquisition of word meaning. In *Proceedings of the Fourth International Conference on Cognitive Modeling*, E. M. Altmann, A. Cleeremans, C. D. Schunn, & W. D. Gray (eds.), 67–72. Mahwah, NJ: Lawrence Erlbaum.
    2002   Modeling the development of the lexicon with a growing self-organizing map. In *Proceedings of the Sixth Joint Conference on Information Science*, H. J. Caulfield et al. (eds.), 553–556. Association for Intelligent Machinery, Inc.

Gathercole, Susan E. and Alan D. Baddeley
  1993   *Working Memory and Language.* Hillsdale, NJ: Erlbaum.
Gupta, Prahlad and Brian MacWhinney
  1997   Vocabulary acquisition and verbal short-term memory: Computational and neural bases. *Brain and Language* 59: 267–333.
Hebb, Donald O.
  1949   *The Organization of Behavior: A Neuropsychological Theory.* New York, NY: Wiley.
Hernandez, Arturo, Ping Li, and Brian MacWhinney
  2005   The emergence of competing modules in bilingualism. *Trends in Cognitive Sciences, 9,* 220-225.
Holland, John H.
  1975   *Adaptation in Natural and Artificial Systems.* Ann Arbor, MI: The University of Michigan Press.
James, Daniel L. and Risto Miikkulainen
  1995   SARDNET: A self-organizing feature map for sequences. In *Advances in Neural Information Processing Systems 7,* (NIPS'94) G. Tesauro, D. S. Touretzky, and T. K. Leen (eds.), 577–584. Cambridge, MA: MIT Press.
Kohonen, Teuvo
  2001   *The Self-Organizing Maps.* 3rd Ed. Berlin: Springer.
Ladefoged, Peter
  1982   *A Course in Phonetics.* 2nd Ed. San Diego, CA: Harcourt Brace.
Li, Ping
  2000   The acquisition of tense-aspect morphology in a self-organizing feature map model. In *Proceedings of the Twenty-Second Annual Conference of the Cognitive Science Society,* Lila R. Gleitman & Aravind K. Joshi (eds.), 304–309. Mahwah, NJ: Lawrence Erlbaum.
  2003   Language acquisition in a self-organizing neural network model. In *Connectionist Models of Development: Developmental Processes in Real and Artificial Neural Networks,* P. Quinlan (ed.), 115–149. Hove & Briton: Psychology Press.
  2006   In search of meaning: The acquisition of semantic structure and morphological systems. In *Cognitive Linguistics Investigations: Across Languages, Fields and Philosophical Boundaries* J. Luchjenbroers (ed.), 109–137. Amsterdam: John Benjamins.
Li, Ping and Melissa Bowerman
  1998   The acquisition of lexical and grammatical aspect in Chinese. *First Language* 18: 311–350.
Li, Ping and Brian MacWhinney
  1996   Cryptotype, overgeneralization, and competition: A connectionist model of the learning of English reversive prefixes. *Connection Science* 8: 3–30.

2002   PatPho: A phonological pattern generator for neural networks. *Behavior Research Methods, Instruments, and Computers* 34: 408–415.

Li, Ping and Yasuhiro Shirai
2000   *The Acquisition of Lexical and Grammatical Aspect*. Berlin/New York: Mouton de Gruyter.

Li, Ping, Curt Burgess, and Kevin Lund
2000   The acquisition of word meaning through global lexical co-occurrences. In *Proceedings of the Thirtieth Stanford Child Language Research Forum*, E. V. Clark (ed.), 167–178. Stanford, CA: Center for the Study of Language and Information.

Li, Ping, Igor Farkas, and Brian MacWhinney
2004   Early lexical development in a self-organizing neural networks. *Neural Networks* 17: 1345–1362.

Li, Ping, Xiaowei Zhao, and Brian MacWhinney
2007   Dynamic self-organization and early lexical development in children. *Cognitive Science* 31: 581–612.

Ling, Charles X. and Marin Marinov
1993   Answering the connectionist challenge. *Cognition* 49: 267–290.

MacWhinney, Brian
1993   Connections and symbols: Closing the gap. *Cognition* 49: 291–296.
1998   Models of the emergence of language. *Annual Review of Psychology* 49: 199–227.
2000   *The CHILDES project: Tools for Analyzing Talk*. Hillsdale, NJ: Lawrence Erlbaum.
2001   Lexicalist connectionism. In *Models of Language Acquisition: Inductive and Deductive Approaches*, P. Broeder & J. M. Murre (eds.), 9–32. Oxford, UK: Oxford University Press.

MacWhinney, Brian and Jared Leinbach
1991   Implementations are not conceptualizations: Revisiting the verb learning model. *Cognition* 40: 121–157.

Marcus, Gary F., Michael Ullman, Steven Pinker, Michelle Hollander, T. John Rosen, and Fei Xu
1992   Overregularization in language acquisition. *Monographs of the Society for Research in Child Development* 57(4): 1–182.

McShane, John and Stephen Whittaker
1988   The encoding of tense and aspect by three- to five-year-old children. *Journal of Experimental Child Psychology* 45: 52–70.

Moens, Marc and Mark Steedman
1988   *Computational Linguistics* 14: 15–28.

Miikkulainen, Risto
1997   Dyslexic and category-specific aphasic impairments in a self-organizing feature map model of the lexicon. *Brain and Language* 59: 334–366.

Olsen, Mari Broman, Amy Weinberg, Jeffrey P. Lilly and John E. Drury
    1998   Mapping innate lexical features to grammatical categories: acquisition of English -*ing* and -*ed*. In *Proceedings of the 20th Annual Conference of the Cognitive Science Society*, M. Gernsbacher and S. Derry (eds.), 794–799. Mahwah, NJ: Lawrence Erlbaum.
Passonneau, Rebecca J.
    1988   A computational model of the semantics of tense and aspect. *Computational Linguistics* 14: 44–60.
Pinker, Steven
    1994   *The Language Instinct: How the Mind Creates Language*. New York, NY: Harper Collins Publishers Inc.
Pinker, Steven and Alan Prince
    1988   On language and connectionism: Analysis of a Parallel Distributed Processing Model of language acquisition. *Cognition* 29: 73–193.
Plunkett, Kim and Virginia A. Marchman
    1991   U-shaped learning and frequency effects in a multilayered perceptron: Implications for child language acquisition. *Cognition* 38: 43–102.
Rogers, Timothy T. and James L. McClelland
    2004   *Semantic Cognition: A Parallel Distributed Processing Approach*. Cambridge, MA: MIT Press.
Ritter, Helge and Teuvo Kohonen
    1989   Self-organizing semantic maps. *Biological Cybernetics* 61: 241–254.
Rumelhart, David E. and James L. McClelland
    1986   On learning the past tense of English verbs. In *Parallel Distributed Processing: Explorations in the Microstructure of Cognition*, J. L. McClelland and D. E. Rumelhart (eds.), 216–271. Cambridge: MIT Press.
Scheler, Gabriele
    1996   Learning the semantics of aspect. In *New Methods in Language Processing*, H. Somers and D. Jones (eds.), 83–95. London: Routledge.
Shirai, Yasuhiro
    1991   Primacy of aspect in language acquisition: Simplified input and prototype. Ph.D. dissertation, Applied Linguistics, University of California, Los Angeles.
Shirai, Yasuhiro and Roger W. Andersen
    1995   The acquisition of tense-aspect morphology: A prototype account. *Language* 71: 743–762.
Shultz, Thomas R.
    2003   *Computational Developmental Psychology*. Cambridge, MA: MIT Press.
Silberman, Yaron, Shlomo Bentin, and Risto Miikkulainen
    2007   Semantic boost on episodic associations: An empirically-based computational model. *Cognitive Science* 31: 645–671.

Siegel, Eric V. and Kathleen R. Mckeown
   2000   Learning methods to combine linguistic indicators: Improving aspectual classification and revealing linguistic insights. *Computational Linguistics* 26: 595–628.
Smith, Carlota S.
   1983   A theory of aspectual choice. *Language* 59: 479–501.
   1997   *The Parameter of Aspect.* Dordrecht: Kluwer.
Vendler, Zeno
   1957   Verbs and times. *Philosophical Review* 66: 143–160.
Zhao, Xiaowei and Ping Li
   2005   A self-organizing connectionist model of early word production. In *Proceedings of the 27th Annual Conference of the Cognitive Science Society*, B. G. Bara, L. Barsalou, & M. Bucciarelli (eds.), 2434–2439. Mahwah, NJ: Lawrence Erlbaum.
   2007   Bilingual lexical representation in a self-organizing neural network. In *Proceedings of the 29th Annual Cognitive Science Society*, D. S. McNamara & J. G. Trafton (eds.), 755–760. Austin, TX: Cognitive Science Society.
   2008   Vocabulary development in English and Chinese: A comparative study with self-organizing neural networks. In *Proceedings of the 30th Annual Conference of the Cognitive Science Society*, B. C. Love, K. McRae, & V. M. Sloutsky (eds.), 1900–1905. Austin, TX: Cognitive Science Society.
   in press   The acquisition of lexical and grammatical aspect in a developmental lexicon model. *Linguistics: An Interdisciplinary Journal of the Language Sciences.*

Contributors

Jürgen Bohnemeyer
Department of Linguistics

University at Buffalo – SUNY
609 Baldy Hall
Buffalo, NY 14260
USA

jb77@buffalo.edu

Mary Carroll
Seminar für Deutsch als Fremdsprachenphilologie

Ruprecht-Karls Universität Heidelberg
Plöck 55
69117 Heidelberg
Germany

carroll@idf.uni-heidelberg.de

Todd R. Ferretti
Department of Psychology

Wilfrid Laurier University
75 University Avenue West
Waterloo, Ontario
Canada N2L 3C5

tferrett@wlu.ca

Wolfgang Klein
Language Acquisition

Max Planck Institute for Psycholinguistics
Wundtlaan 1
6525 XD Nijmegen
The Netherlands

wolfgang.klein@mpi.nl

Ping Li
Department of Psychology

Pennsylvania State University
University Park, PA 16802
USA

pul8@psu.edu

Carol J. Madden
Institute of Psychology

Erasmus University Rotterdam
Burgemeester Oudlaan 50
3062 PA Rotterdam
The Netherlands

madden@fsw.eur.nl

Yasuhiro Shirai
Department of Linguistics

University of Pittsburgh
2816 Cathedral of Learning
Pittsburgh, PA 15260
USA

yshirai@pitt.edu

Arnim von Stechow
Seminar für Sprachwissenschaft

Universität Tübingen
Wilhelmstr. 19–23
72074 Tübingen
Germany

arnim.stechow@me.com

Christiane von Stutterheim
Seminar für Deutsch als Fremdsprachenphilologie

Ruprecht-Karls Universität Heidelberg
Plöck 55
69117 Heidelberg
Germany

cvs@idf.uni-heidelberg.de

Xiaowei Zhao
Department of Psychology

University of Richmond
Richmond, VA 23173
USA

xzhao2@richmond.edu

Index

absolute tense (=deictic tense), 2, 47, 84, 102, 112, 113, 159, 161
accomplishment, 58, 60, 63, 89, 131–132, 136–138, 173–175, 177, 180, 218–221, 229, 245–249, 258
achievement, 58, 60, 63, 78, 89–90, 134, 131–138, 173–177, 180, 182, 230, 245–246, 249, 258
activity, 17, 58–64, 76, 88–89, 132, 136–138, 172–177, 184, 195, 202, 204, 220, 228, 230, 241–242, 245–249, 253, 258–264
adverbial (temporal), 1, 6, 25–28, 33–35, 40–43, 47, 53, 59, 62–71, 77, 84–86, 91, 99–100, 109–114, 119–122, 169, 195, 201, 205, 207, 217, 247
Aktionsart, 18, 40–42, 59, 77, 132, 211
anaphoric, 32–35, 65, 83–85, 102, 108, 112, 157, 162
anaphoric tense (=relative tense), 47, 84, 102, 110, 113, 140, 142
anthropology, 6, 21
arrow of time, 12, 21
aspect, 1–4, 16, 18, 40–43, 52–59, 62–63, 69–71, 77–78, 87, 92–93, 95, 107, 109, 111, 115–116, 119, 123, 168–169, 172–185, 197–199, 202, 205, 214, 217–237, 241–251, 259–266
Aspect First Hypothesis, 173, 235
Aspectual Classification, 220, 236, 246–248, 265
atelic, 62, 90–91, 174, 177, 179–180, 219–220, 229, 232, 245, 247, 249–250

binding implicature, 119–120, 122
biology, 2, 6–7
bound tense, 155, 159
boundary, 25, 33, 52, 57–64, 98, 198, 206–207, 222

calendaric, 26, 32–33, 49, 65, 69
change, 2, 5, 11–22, 28–29, 34, 55–56, 60, 62–63, 88–90, 134, 137, 172, 174, 195, 199, 202, 206–207, 211–214, 219–220, 228, 242, 247, 255, 264
Chinese, 1, 6, 17, 40–41, 70, 168, 172–173, 178, 183–184, 196, 198
coding time (=utterance time), 33, 43, 67, 84–85, 99, 102, 104–107, 112–114, 117, 119, 122
Computational Model, 4, 223, 241–246, 249, 265–266
connective (temporal), 83–87, 100–102, 107, 115, 119–120
culture, 5, 16–18, 21, 25–26
cyclic time, 5, 16, 21

deictic, 2, 28–29, 32–35, 43, 47, 49, 56, 65, 83–85, 102, 108, 117, 119, 122, 131, 140, 142, 146, 152, 154–156, 159, 161, 174
default past tense hypothesis, 182–183
DevLex-II model, 251–254, 258, 265
discourse principle, 1–2, 35, 41, 70
Discourse Representation Theory (DRT), 83–85, 116, 121
distributional bias hypothesis, 176–177
duration, 5, 9–18, 22–23, 27–30, 41, 60, 61, 63, 66–69, 73, 91, 99, 131, 169, 176, 213–214, 217–220, 223–224, 230, 233–236, 245
Dutch, 199, 205–214

Emergentism, 171, 174, 176–177, 242, 244, 250, 260, 262, 264–266
empirical research, 218, 231, 236
endpoint, 57, 205, 207–210, 213, 217, 219–220, 224–226, 229–230, 233, 245, 247, 259

English, 1–3, 5, 26, 36, 39–45, 50–57, 62–68, 74, 77–78, 83, 85, 89, 101–102, 112–116, 119, 123, 130, 140–143, 152–153, 162–163, 169–177, 180–184, 195–201, 205–212, 221–222, 234, 242–248, 255–256, 259, 265
epic preterite, 50
event, 1–3, 5, 8, 10–12, 15–17, 22–23, 26, 28–35, 39–42, 47–52, 57–61, 65, 70–72, 84–85, 88–90, 93, 95–102, 107–123, 130, 137–138, 147, 149–150, 153, 155, 168, 172, 174, 181, 183, 195–199, 203–214, 217–221, 229, 233–234, 241, 244–248
event construal, 199
event realization, 95, 109, 111, 117
eye tracking, 208–210

first language acquisition, 4, 168–169, 173
frequency, 41, 66–69, 73, 86, 169, 177–178, 247, 266
future, 1–18, 23–25, 28, 32, 43–45, 49, 51, 55, 69–70, 83, 87, 93–99, 102, 104–105, 107–111, 113, 122, 142, 150–153, 158–162, 171, 184–185, 196, 218, 232–233, 235–236, 241

German, 11, 41, 44–45, 47, 50, 67, 123, 143, 172, 184, 199–200, 205–214
god, 6, 10, 217

Hebbian Learning, 4, 250–254, 257, 265–266

iconicity, 120
imperfective, 40, 42, 52–58, 63, 76–78, 84, 87, 91–96, 103, 107, 111, 116–118, 173, 183–184, 197–198, 214, 220–236, 245, 248, 259
implicature, 85, 95, 101, 113, 119–122
inclusion, 27, 95, 117–118
innateness, 4, 175, 249

input, 175–178, 182, 198, 242–243, 248–262, 265–266
irrealis mood, 87

Kalaallisut (West Greenlandic), 83, 109, 116, 121–122

language, 1–7, 16–18, 25–29, 32–36, 39–46, 51–67, 70, 72, 75, 78, 83–91, 113–123, 130–132, 138–139, 167–185, 195–201, 205–206, 209–211, 214, 217–218, 228–229, 232–237, 241–244, 246, 248–251, 265–266
Language Bioprogram, 175, 249
lexical, 1–4, 7, 18, 26, 30, 40, 52, 55–61, 64–65, 74–78, 84, 86, 92–99, 102, 123, 132, 134–137, 139–140, 146, 149–150, 155, 167–176, 179–182, 196, 200–204, 229–232, 245–251, 255, 257, 259, 260–266
lexical aspect, 2, 4, 18, 40, 52, 59, 78, 171–176, 179, 180–182, 201, 245–251, 257, 259, 261–266
lexical Representation, 255
LF, 133, 138–145, 148–150, 154, 157–158, 163
linear time, 16–17, 21

mental representation, 3–4, 217–218, 223, 229, 231–232, 235
metrical tense, 99, 110, 112
modal commitment constraint, 109–110
multiple-factor account, 177

narrative present, 49
natural temporal reference point, 101, 117–122
Neural Network, 177, 242–244, 248, 249–251, 254, 265–266
now, 1–2, 5–6, 13–15, 19, 23–30, 33–34, 40–51, 58, 62–65, 74, 108, 113, 118, 132–133, 137, 139, 142–143, 148, 156, 158, 160, 168–170, 173–174, 179, 202–206, 232–234, 249

Index 277

ongoing, 69, 70, 116, 177, 218, 221, 223–230, 233–236, 241, 245, 259
ontology of time, 130
order, 1, 8–18, 21, 23, 27–31, 41, 47, 52, 62, 71–72, 101, 114, 117, 138, 142, 151–152, 156, 168, 174, 201, 208–212, 225, 236, 244, 253–254
origo, 28–30, 33–34, 130–131

Partee problem, 130
particle, 1–2, 41, 69, 70, 87, 100
past, 1, 3, 5–6, 8–10, 12, 14–17, 23–25, 28, 34, 42–55, 65–71, 74–76, 87, 93–98, 101–110, 122, 129, 140–162, 169–184, 196, 222, 225, 227, 233–235, 241–244, 257, 259, 265
perfect, 53–54, 70, 77, 96, 116, 141–143, 146–147, 152, 159, 197, 220, 222–224, 227–229, 231, 236
perfective, 40, 43, 52–58, 84, 87, 92–96, 101–102, 107–110, 115, 117–122, 147, 149–152, 173, 175–178, 180, 182–184, 197–198, 220, 223–236, 245, 250–259, 266
phase, 60, 89, 91, 206–207, 209–210, 220, 245
philosophy, 7–8
physics, 2, 6–7, 9–13, 21–23
praesens tabulare, 50
preferred topic time selection, 118
present, 1–2, 4, 5–17, 23–25, 28, 31–34, 40–59, 68–69, 74, 83–84, 91, 93, 102, 104, 108, 113, 116, 118, 122, 129–130, 137, 140–148, 151–156, 159–162, 167–168, 171, 176, 184, 195–198, 202, 210–211, 214, 218, 222, 233–237, 241, 261
present perfect, 53, 68, 141, 146, 151, 234–235
Present Perfect Puzzle, 146
principle of chronological order, 35, 72
process, 3–5, 8, 12–14, 16, 22, 39, 52, 58–60, 76, 88, 99, 170, 175, 184, 195, 198, 210, 217–223, 229–237, 241, 244, 247–254, 257, 259–261, 265–266

progressive, 53–54, 56, 62, 85, 87, 94, 96, 103, 107, 111, 115–119, 122, 142, 149, 156, 172–184, 197–198, 204–205, 213–214, 219–222, 225, 233, 245–246, 248–250, 254, 259, 265–266
prototype, 176–184, 222, 236
proximity, 28–29, 98
psychology, 2, 6–7, 14, 24, 218, 223–224, 236, 242

realis mood, 93
reality, 8, 12, 19, 21, 42, 48, 60–61, 95, 195
reference point, 84, 88, 99–102, 108, 112–113, 117–122
referential shift, 119–120
region, 29–32, 233, 253, 262–263
regular-irregular debate, 171
relativity, 11–12
relatum, 31–35, 65
Russian, 40–43, 52–56, 58, 123, 152, 172, 197–198, 248

second language acquisition, 3, 167, 169, 172–173, 182
segmentability, 27
Self-Organizing Map (SOM), 250, 252–253, 256
simultaneous, 9, 15, 23, 43, 45–46, 51, 65, 71, 83, 101, 121, 123, 152–155
situation, 3–5, 18, 20, 25, 32–35, 39–53, 56–76, 84, 97, 133, 174, 176–177, 180, 183, 195, 198, 201–214, 217–237, 241, 245–246
specific language impairment (SLI), 173, 178–179
speech onset times, 208–209
state, 1–3, 5, 12–15, 18, 21, 26, 29, 39–40, 42, 58–66, 67, 74–75, 84–85, 88–89, 99, 107–110, 113, 115, 117–122, 132–138, 172, 174, 177–182, 198, 203–204, 206, 210–211, 213–214, 218–223, 226, 234, 236, 245–247, 264
status inflection, 86–87, 91

subordinate tense, 152
succession, 8–11, 15–18, 23, 26–27, 160

telic, 62, 77, 90–91, 174–175, 177, 179, 180, 182, 184, 219–220, 232, 245–250, 258–259, 262–264
temporal adverbs, 1–2, 18, 40–41, 49, 57, 62, 64–67, 76, 99, 114, 129–130, 146–147, 151, 159–160, 168–169, 199, 246
temporal anaphora, 2–3, 83–85, 102, 113–114, 116–117, 121–122
temporal auxiliaries, 130, 150, 162
temporal PRO, 3, 27, 40, 62, 72–73, 132, 153, 158–159, 162, 211, 218, 220, 247
tempus, 6–7, 9, 43
tense, 1–2, 5, 7, 18, 25, 28, 32–34, 40–59, 65, 68–71, 77–78, 83, 85, 96, 108, 110, 113–116, 123, 129–132, 137, 140–145, 149–163, 168–196, 199, 217–218, 232–235, 237, 241–244, 257, 265–266
tense deletion, 159, 161
tenselessness (tenseless languages), 2, 83, 85, 102, 109, 112–117, 132, 140, 153, 163, 168

time, 1–36, 39–78, 83–85, 88, 93–104, 107–122, 129–162, 168–170, 180, 183–184, 195–199, 202–207, 217–220, 228, 230–234, 241–246, 250, 253–255, 259
time argument, 130, 132, 135, 137–138, 140–141
topic time, 46, 51, 66, 76, 78, 83–85, 94–122, 149, 204, 206

Vendler Aktionsarten, 130, 132
verb class, 61, 63, 201, 218–220, 222–223, 229, 231–232, 236–237
Verb Meaning, 204, 264
viewpoint, 40, 52–58, 84, 86, 93, 101, 119, 123, 205, 210–212, 245
viewpoint aspect, 52, 86, 101, 119, 123, 245

Yucatec (Maya), 2, 83–94, 100, 102, 110, 113–123